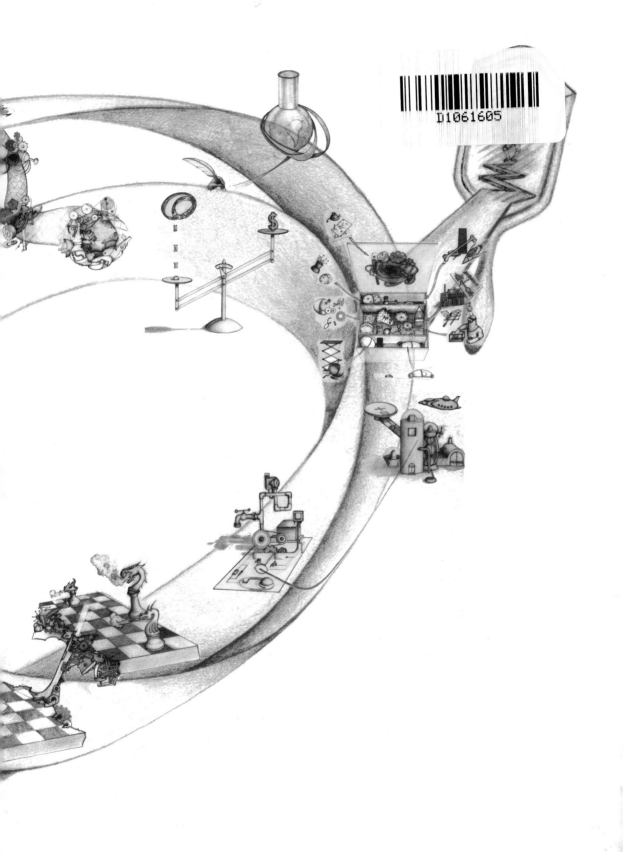

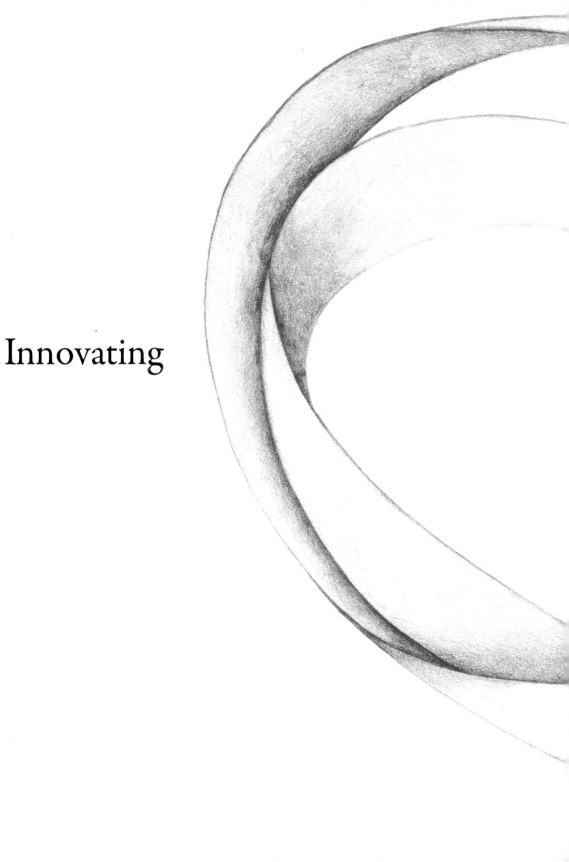

Innovating

Innovating

A Doer's Manifesto
for Starting from a Hunch,
Prototyping Problems,
Scaling Up, and Learning
to Be Productively Wrong

Luis Perez-Breva

Artwork by Nick Fuhrer

Foreword by Edward Roberts

THE MIT PRESS   CAMBRIDGE, MASSACHUSETTS   LONDON, ENGLAND

This book was set in Garamond and Trade Gothic by the MIT Press. Printed and bound in the United States of America.

Library of Congress Cataloging-in-Publication Data

Names: Perez-Breva, Luis, author.
Title: Innovating : a doer's manifesto for starting from a hunch, prototyping
  problems, scaling up, and learning to be productively wrong / Luis
  Perez-Breva ; foreword by Edward Roberts.
Description: Cambridge, MA : MIT Press, [2016] | Includes bibliographical
  references and index.
Identifiers: LCCN 2016017241 | ISBN 9780262035354 (hardcover : alk. paper)
Subjects: LCSH: Creative ability in business. | Problem solving. | New
  products. | Technological innovations. | Entrepreneurship.
Classification: LCC HD53 .P4694 2016 | DDC 658.4/063--dc23 LC record available
at https://lccn.loc.gov/2016017241

10   9   8   7   6   5   4

to Erin, to Owen, to Marta ... my board of directors

CONTENTS

FOREWORD
Edward Roberts

I've spent more than 50 years building entrepreneurship and innovation research and teaching at MIT, combined with countless experiences founding, advising, and investing in many tens of new and growing enterprises— and I find every page of *Innovating* a treasure trove of fascinating new perspectives and insights. Just look at the subtitle of this wonderful book about doing Luis Perez-Breva has created: "A Doer's Manifesto for Starting from a Hunch, Prototyping Problems, Scaling Up, and Learning to Be Productively Wrong." Other than "scaling up"—which, of course, Luis treats differently than you may be used to (including even talking about "scaling down")— have you ever seen those words used to frame a treatise on how to innovate?

Let's start from the underlying thrust of the book: What is the knowledge or evidence one needs to make a problem real, find one or more appropriate and effective solutions, and bring a chosen solution forward into practice? Luis tells us to begin with a "hunch." Wait a minute!! This is MIT!!! Isn't he going to expound upon the scientific basis for innovating? Isn't he going to lay out a series of formulaic steps to make us the fortunes we seek, or at least get us to our desired solutions? No, "starting with a hunch" is where readers will begin to sense that they are in for a very different learning experience.

Think about "hunches": a hunch about what might be a problem you really want to be tackling; a hunch about approaches that might help in solving a problem; a hunch about the "parts" (his word) that would go into a search for a solution and into the solution itself; a hunch about the kinds of people you will need as you make progress at each stage; a hunch about when and how to carry out your possible (eventual?) fundraising, and so on. Luis doesn't confront you with some bold answer to your needs. Instead, he challenges you to recognize how much you don't know at each phase of your innovating activity and how much (valuable) uncertainty there is as you move forward. And then he helps you steadily to understand better how to proceed in your search and decide whether you want to move ahead—and if so, how. The path to impact is fraught with near misses and learning.

So, here's a guy who doesn't follow the path of most books on entrepreneuring (there aren't many books on *how* to innovate!) in which the author exhorts you to follow his revealed wisdom as he tells you the "secret" of how he made his first hundred million or his first billion dollars! Instead, Luis—a very experienced and pragmatically focused teacher—shows you how he has brought students and people in companies of all ages and with different levels of experience and knowledge further along in their quest to solve problems. And if you are starting from just a hunch, then clearly you should expect you will often be wrong—which is fine.

The book proceeds from that kind of thought process to become an aid to your search for solutions to real-world problems, whether product or process or social issue or organizational design. The author takes the same approach to all of these, and repeatedly offers examples selected from that diversity to help you better understand how to innovate in many contexts.

An underlying "philosophy" (my label) throughout this book is modesty about how much you can know when you are trying to do something that you want to be different from what exists. Luis says that in reality both the problem itself and its solution are often discovered as part of the innovating process. That is a far cry from providing a formula needing only plug-ins. And in his advice on the process to follow, he draws many non-obvious conclusions. For example, if the "real problem" you first identified seems too big to tackle immediately, perhaps you need to formulate your problem at a smaller scale than you had imagined so you can start working on that lesser reality.

One big thing Luis continually asks of you is to reexamine what you believe is the logical next step, using simple words that need defining in the context (which he does for you). What are "parts" of a problem and its prospective solution? They are, of course, the list of all those elements that show up in a piecemeal taking apart of a device you might want to build, but also those elements of an organization you might need to construct, and as well of the assembly of people you might require to determine a solution and then implement it. And even then, more details are desirable. For example, there are the "sub-parts" people bring in— information, capabilities, and skills; these, too, become parts for specific consideration. Luis wants you to be explicit about all of these, and encourages you to seek out as many of these parts in simple, nearby, approximate and inexpensive forms, to help you move along with what you are trying to do. It is again the modesty of manner that permits you to go forward with little resources and with continuity.

Luis is himself extremely innovative in presenting the concept of "innovation prototyping" and elaborating how to construct a "kit" for that process. He writes that it will have "a *hunch* dressed as a problem, *a set of accessible parts*, *pointers to impact*, *pointers to people*, and a *primer* on how to work on the parts and the impact *at scale*." Every one of those highlighted concepts (my emphasis) demonstrates a toe-in-the-water starting approach for eventually accomplishing a significant goal.

The author reminds you frequently that your purpose is always to find out how you are wrong, so you can then think about a proper next step of inquiry and action. He remarks that "failing sooner buys you time, and money!" while insisting that, of course, you will "fail" (be wrong? make mistakes?) in every stage of your endeavor. And he often does this with wholly unanticipated uniqueness. His presentation of "Fundraising as Advocacy" in a ballet in three movements (!) is an outrageous and absolutely fascinating way to communicate a process that is so awkward and time consuming and yet essential to all forms of innovating.

Let me comment about my experience of reading the book (and you'll no doubt go back and reread some of the chapters more than once to be sure you've got it right). Each chapter has multiple boxes that illustrate and expand the message of the main text. The examples in the boxes range over fields of human endeavor from creativity to science to inventions to industrial and technological history. Their breadth brings home the points Luis is making. Each chapter has highlighted quotes as you move along to help you capture the essence of the arguments being made. Each chapter ends with "takeaways" that synopsize the main points you should have understood. They prod you to return and reread some part of the chapter that in afterthought seems important enough to reexamine. Each chapter also ends with a splendid illustration that creatively integrates all the major considerations of the chapter. The artwork is quite imaginative, and some readers will find the illustrations to be further stimulants of their thinking. The epilogue provides academic commentary that directs you to underlying data, to deeper reading opportunities, and to further explanations.

If you want, you can just read the takeaways at the end of all the chapters. They will give you continuity and will clearly present the essence of the book's uniqueness of thought. You would get a boatload of ideas and insights. But you will have missed the enormous content about doing, and the subtleties that no doubt will make the difference in whether you can really follow Luis Perez-Breva's own guidance to your innovating success: "... bringing together parts, people, and scale in a way not done before [is] something you do continually; innovation is the afterthought."

I hope you enjoy this book as much as I did.

Edward Roberts is David Sarnoff Professor of Management of Technology, Sloan School of Management, Massachusetts Institute of Technology; Founder and Chair, MIT Entrepreneurship Center (now named the Martin Trust Center for MIT Entrepreneurship)

ACKNOWLEDGMENTS

The story of how this book came to be begins in Fall 2012 with a comment from my friend and mentor, Professor Charles Cooney of MIT: "I think you may have a book." I have Charlie to thank for a lot, from giving me the opportunity back in 2007 to begin building innovating as a new field to seeding the hunch that gave rise to this book.

Professor Edward Roberts of MIT had an overwhelmingly positive reaction after reading the manuscript and wrote a foreword that is a far better endorsement of my work than I could ever have hoped for. I am deeply grateful for Ed's caring and straightforward mentoring, and for the knowledge that I can always trust Ed to tell me when he thinks I am wrong—the attribute I value most in people.

I have many students, colleagues, and friends to thank for the opportunity to develop this field by "doing." I must single out the opportunity I was given to develop a month-long Innovation Workshop—evolved from collaborations across the globe—for the new Skolkovo Institute of Science and Technology. It was Skoltech's first course offering ever (given at MIT) and has since become the entry experience for all incoming students on the campus in Russia.

Worldwide, more than 3,000 individuals ranging from undergraduates to senior executives have experienced different aspects of the book's content through my teaching. Their sharp questions and insights have made me better. I am indebted to them all for keeping me honest.

I knew nothing about writing a book. Following the "hunch" (a central concept in this book) that grew out of Charlie's comment, and my own advice, I "sought out people" (another central concept) who knew far more about writing books than me. I found many. I am especially grateful to Gita Manaktala and John Covell of the MIT Press for educating me and supporting me on the steps from that hunch to book proposal over a two-year period. Among other MIT Press staff, I am particularly indebted to Emily Taber—who took over as my acquisitions editor when John retired—for making everything happen.

Many others helped me along that process with support, advice, and sometimes patient tolerance of my inclination for hyperbole and finding a path for every idea to evolve into an absurd plan for world domination (you know who this is for): Prof. Duane Boning, Maren Cattonar, Elizabeth Cooper, Kelly Courtney, Daniela Couto, Prof. Ed Crawley, Ilia Dubinsky, Jose Estabil, Gadi Geiger, Bryan Haslam, Prof. Douglas Hart, Tylor Hess, Christopher Holland, Winston Larson, Marilyn Levine, Katey Lo, Mariana Matus, Vicente Montes, Lisa Natkin, Andrew "Ozz" Oswald, Amadeo Petitbò, Julian Rodriguez, Rosangela dos Santos, Harry Schechter, Jim Schumacher, Patrice Selles, Prof. Maurizio Sobrero, Diane Soderholm, Prof. Bruce Tidor, and Rebecca Walsh. Not helpful was Ignatius J. Reilly.

I must also acknowledge MIT for giving me the opportunity to deploy in classes and workshops the thinking that underlies this book, and particularly to evolve iTeams into an innovation reactor from which we've steered more than 150 MIT technologies to impact. The challenges of navigating the strong disciplinary silos at MIT steeled my resolve to advance with my ideas and this book project. *Future-me* will be thankful for that.

Two people deserve a special mention:

Scott Cooper has been my editor; I am most grateful for his dedication, astonishing attention to detail, patience, and explanations. I had no real idea what an editor did before I endeavored to write a book, and I'm still not sure, but whatever an editor is supposed to do, Scott did more: He joined me with tremendous enthusiasm, pulled no punches, taught me how to write better, and worked tirelessly with me to wield the English language into delivering an unassuming and forward-looking prose that makes the writing as direct as I perceive the content to be. Then he handled many things with the publisher one would expect of an agent. Thank you, Scott; it has been a lot of fun (the antecedent of it here is at once both "the covenant" and "future English").

Nick Fuhrer helped me realize a vision I was told by many could not be achieved ("that's not how these books are supposed to look"). His detailed artwork is nothing short of exquisite. I wanted to offer readers' imaginative and creative side an experience that traditional diagrams (whether technical or simplistic) and cartoon representations simply can't provide. I am convinced Nick summoned the spirits of Escher and Dalí to produce the dynamic artwork that will surely help many appreciate innovating as the highly rewarding, industrious, yet accessible process it is. I am indebted to Nick for helping me realize my vision, and for making it so easy to work and translate the concepts I visualized into exquisite art. I am looking forward to find an excuse to work together again.

One last thing about my collaboration with Nick: I also knew nothing about talking to illustrators or even how to find one. For as long as I tried to find an "illustrator," I was unable even to define what I thought I needed. At some point in the future, in hindsight (another important concept in the book), I'll gloss over the full story of how I found Nick and will say I found an illustrator. But that's not how it happened. The acknowledgements section is not the place to begin developing the content, so suffice it to say that the beginning of our collaboration benefited from the same kind of happenstance I suggest readers embrace (chapter 4). Nick introduced himself as a sculptor, I spoke of myself as a doer, and it turned out we knew each other already. But it took a third person and a casual encounter at the Shady Hill School in Cambridge, Mass., for us to connect the dots. Because of that, I am also grateful to the Shady Hill School community.

While they did not participate directly in this project, I would also like to acknowledge Noubar Afeyan and Fiona Murray. The conversations we have had over the years in our various collaborations have made me a better thinker. I also thank the following people for having believed in me every time I transgressed a disciplinary boundary: Manlio Allegra, Michel Brune, Tommi Jaakkola, Enric Julià, Luis Piera, Tommy Poggio, and Ken Zolot.

Like many of you doers, I did not have the luxury of halting everything else to write this book. Along the process, I gave up sleep and comfort to find the time. I can't imagine succeeding without the help from Susana, Pedro, and Maria Jose, and without the support of my parents Pilar and Jose Antonio and my brothers Jose Antonio and Manuel.

Finally, three people made this book possible above all others in the three years that followed the hunch that began the project. My daughter Erin graciously conceded the time we typically devote to building contraptions and doing experiments. My son Owen has had to wait to join us in our crazy inventions. And my wife Marta has always been there for me. I can't imagine life without her, her unwavering support, sharp insights, and straightforward advice. Through the duration of this project she was also our family's only practicing entrepreneur. That she managed to keep us all afloat while she effectively executed her senior management responsibilities in a biotech startup is nothing short of awe-inspiring. More important than what Marta, Erin, and Owen gave up is the enthusiasm with which they carried me through the entire project. They are my family and board of directors. And each one of them got to decide something about what you are about to experience. They deserve this prime spot in my acknowledgments.

INTRODUCTION

I introduce innovation and entrepreneurship to a wide variety of audiences, including university students at the undergraduate and graduate levels and executives in companies. There's one thing I can almost always count on: The audiences have already been taught to see innovation and entrepreneurship as one and the same. And so, as they try to apply their skill sets to the subject matter, the same contradictions, paradoxes, and even sense of frustration seem to kick in—no matter their backgrounds.

I attribute this to two fundamental things. The first is an overabundance of so-called recipes for creating a startup or having an innovation. Students come to class hoping I will give them a subroutine that they will merely have to execute, as if they were computers. Many have been led to identify any experience with or instruction in preparing to "pitch" a business concept to others with actual preparation for *conceiving* a solid idea for innovation or *executing* on a business concept. The second is a general lack of acknowledgment that the very same language that can be so powerful for articulating a business, developing a business strategy, and executing that strategy can just as easily mislead the aspiring innovator—especially if applied too soon to what generally amounts to a hunch.

After years of witnessing students making those mistakes, I have come to understand that this situation flows from a conflation of entrepreneurship and innovation—especially in academia. That conflation has led to the creation of a bunch of truisms that focus on managing innovations and organizations, which may be quite useful at some point, but have little or nothing to do with actually innovating. Put simply, we talk a lot about how innovations happen and are managed, but rarely discuss how you actually produce one—a huge multidisciplinary space that is rarely explored. The tasks may require you to venture a bit into the impossible. Being empirical and experimental in that space—something that may seem out of fashion but is sorely missing—is the subject of this book.

The field of entrepreneurship and innovation as we know it is full of amazing stories, inspiring trajectories, and powerful figures—ones we know of only in hindsight. The field is also ripe with opportunities to enter contests and win awards. These "idea" events are a lot like beauty pageants; the pitches are the "talent competition." They sometimes even propel forward aspiring entrepreneurs who have a solid, powerful idea. The key, though, is how solid and powerful their idea was before they entered the event. The pitch itself is show and tell.

Beauty pageants end with the crowning of a winner. Crowned or not, you still need to figure out what you're going to do next.

As people prepare for these events centered on entrepreneurial "beauty," they feel the urge to copy "beautiful"-looking, successful entrepreneurs. One can only wonder how many black turtlenecks, jeans, and wireless headset microphones are sold in advance of these events, or how many prospective presenters practice the line "and we launch today" in front of mirrors. But far more often than not, what we think we know about already successful entrepreneurs includes very little about how they developed their ideas, how they assembled and managed their organizations, or the struggles they went through to evolve their ideas toward impact at scale.

After all, identifying yourself as the "founder" of a business is as easy as paying a state's incorporation fee.

In class, I have learned to help students recognize these contradictions and paradoxes by taking this archetypical conception of entrepreneurship and innovation to its comical extreme: I suggest they incorporate and then post their new status as "founders" on the social network of their choice.

Sometimes I "knight" students—imaginary sword and all—as "recognized entrepreneurs." I give others "official permission" to innovate. I encourage them to find a suitable entrepreneurial beauty pageant to enter. I congratulate them on their success, and then I advise it might be a good time to figure out what their new companies actually do.

There's nothing inherently wrong with the entrepreneurship theory from which my students have developed their paradoxes and contradictions. It's just that the theory is all about managing an organization that has already settled on its target audience(s).

An aspiring entrepreneur or innovator lives at $N = 1$. The merits of her or his innovation or organization will be measured relative to adoption, not by comparison with other innovators or entrepreneurs. The one problem that gives her or him purpose has to be solved with the resources at hand and at scale—it all needs to work. It does not really matter whether the way the problem is ultimately solved falls at the center of some graphical distribution of entrepreneurial performance or at the graph's tail end.

The statistics pertaining to who entrepreneurs are or how they perform do not really apply. That is a limitation of statistics as the chosen method, not a problem with the underlying research. The keywords and concepts used to map entrepreneurial ideas are indexed by the final outcome—a successful startup, a product, an innovation, an enterprise, or more generally the establishment of any kind of organization—not by the initial premise, knowledge, and resources of the entrepreneurs studied.

Keywords and highly specialized concepts such as need, product, distribution, value chain, users, lead users, competitive forces, value creation, and value capture do not have meanings set in stone. At the beginning of an innovator's inquiry, they are largely undefined and ambiguous; they acquire their precise meanings and their analytical strength only over time through the inquiry of the innovator, from the organization that emerges, and in the context of the problem that organization ultimately solves. It's like thermodynamics: We don't need to understand the science to enjoy an iced beverage, but if we ever need to maintain temperature constant for a brief while, the knowledge that temperature remains constant during the transition from liquid to solid may be critical.

It is easy for aspiring entrepreneurs to characterize their ideas using their best understanding of those concepts in the abstract. It is more difficult for them to realize that whatever they end up with may walk and quack like a startup but not yet be a startup. The business concept they may produce remains a good aspirational destination to guide their inquiry, but that's all. I encounter this time and again in class: Incipient entrepreneurs confuse their initial guess of a destination with an actual plan of action.

Unfortunately, it's easy to fall in love with the craft that goes into articulating a concept using precise technical management terms while losing sight of the job ahead. It's the same as burying yourself in technical jargon from whatever field you've been working in. Both are excellent examples of over-engineering—something every engineer is strongly encouraged to avoid.

This is not a shortcoming of the literature of management or that of product design. It is a sign that other fields of inquiry—particularly those concerned with engineering, with high technology, with science, with tinkering, and more generally with the synthesis of new ideas—have yet to offer viable strategies for you to engage in entrepreneurship and innovation that are compatible with that world view. In a way, entrepreneurship and innovation emerged first as a scientific and management field, but they still lack an experimental and engineering footing. Chemistry went through this same process before chemical engineering emerged. A symptom of this lack is that we see more people concerned with idea selection than we see people concerned with actually producing innovations.

The real impact of this shortcoming is that more and more aspiring entrepreneurs and innovators focus on new consumer products and on leveraging reasonably commoditized technologies (e.g., the Web and apps). Meanwhile, fewer pay attention to opportunities in more complex systems and new technologies or use either to conceive entirely new categories of activity. They also fail to address meaningfully how to scale up their ideas until they become viable business concepts to which they could then apply what they have learned (or can learn) about management and entrepreneurship.

This situation persists because the literature and the lessons aspiring entrepreneurs and innovators are applying are intrinsically analytical and statistical and so are most conducive to identifying arbitrage opportunities in well-outlined industries centered on well-identified markets or users. The "toolbox" is biased toward the analysis of what already exists. If an aspiring entrepreneur wants to use the same tools to conceive a new market, to discover an actual real-world problem, or to untangle the complexity of an industry to reveal new opportunities, he or she may discover that the tools demand a significant dose of creativity just to overcome that "bias." That's creativity that is not directly applied to innovating but to make recipes work for something other than what they were intended. We might as well equip aspiring entrepreneurs with broader knowledge about producing innovations so they can channel that same creativity more effectively.

Again, an aspiring entrepreneur or innovator lives at $N=1$, and where he or she lands in a distribution of innovators is immaterial. That isn't the objective. What an aspiring entrepreneur or innovator needs to do is synthesize *one* robust idea—a space of opportunity—and make that work.

A few words regarding this book's tone are warranted.

Much of the language used to describe innovation concepts is contaminated by knowledge of the end points of innovation stories. The contamination renders these concepts useless, even if they are accurate and useful for the analysis of entrepreneurship and innovation stories in hindsight. Entrepreneurs and innovators, however, operate in a highly dynamic environment. That makes static concepts difficult to apply. I see the contamination manifested in some questions from students that really boil down to this: "Would you please now give me the solution at the end of the book?"

I see the same problem elsewhere. Knowing the specific mathematical formula or model constructed to summarize a specific piece of knowledge does not immediately translate into understanding the phenomena that underlie that formula or model. Just as an innovation story has an end point, that formula is an endpoint. Students can apply it skillfully without ever learning to recognize the wide variety of situations in which the original knowledge might apply.

An innovator can rapidly make lots of guesses about business model, value, value proposition, user, and product and get to a semblance of a new venture. Once there, though, it becomes inordinately difficult to unbundle the guesses. The concepts are correct, but if used too early they may fool innovators into mistaking their guesses and the structure built around them for actual evidence of an opportunity. Worse yet, the same tools that result in a guess about an opportunity may not be the tools needed to unbundle that opportunity for the purpose of further experimentation. So, in this book I avoid this "contaminated" language in early chapters and focus instead on an approach to synthesize solutions to real-world problems. I offer strategies for connecting the results of the readers' own inquiries to those concepts in later chapters, after readers have a strong basis upon which to build and evolve their ideas and thus are more likely to apply and use those words in the ways they were originally intended. As the reader's inquiry into a problem progresses, using those words will become critical to adding the last layer of detail to whatever "innovation" is proposed.

The subject of entrepreneurship and innovation is tightly linked to a promise of economic growth, generally through the development of an organization. That strong connection may make my decision to avoid the "contaminated" language I mention above seem odd, and may invite criticism. My choice stems from an observation: Whatever the "innovation endeavor" my students engage in after we first meet, every endeavor is best characterized by everything there was left to learn about the problem they wanted to solve than by any disciplinary technique they brought to it at the outset—no matter how skilled they may be. This choice likely reflects my bias: I understand innovation better as an outcome of an industrious learning process that cannot be fully comprehended from the safety afforded by the methodologies of any one discipline alone. I want readers to enjoy innovating as a learning process, one that can be practiced and that benefits from multiple disciplines and viewpoints. The basis for this motivation is my own experience innovating, and teaching many others to innovate, on various levels and across multiple disciplinary domains.

———————————

Shying away from technical jargon allows me to begin the story for this book at the very beginning of innovating, when everything amounts to a hunch. The chapters in the book follow a sequence that is consequent with that choice, building the concepts for innovating from the perspective of an unassuming doer.

That said, each chapter is written so it can also be read for future reference independent of other chapters, or in a newly created chapter sequence— that is, different than as published here—to suit the reader's specific purposes or "beginning." Reading chapters out of sequence can reveal different perspectives on innovating. For instance, beginning with chapter 6 and then reading chapters 11 and 12 reveals a story about how to implement this approach to innovating as a process in innovation management. Building on that sequence, the remaining chapters provide an innovation manager with specific strategies to help innovators progress through the process from hunch to decision.

The "takeaways" at the end of chapters should make it easy to develop your own sorting function. Cross-references to other chapters should help you design your own path through the book. The academic commentary in the book's epilogue relates the concepts discussed in the chapters to the multidisciplinary literature upon which they rest.

My hope is that, after reading a few chapters, you'll become curious about the subtle shift in mind-set regarding innovating that's portrayed in the book and you'll feel free to read the rest of the chapters in an order that suits your interests. Feel free to do so, just as *Star Wars* fans have come up with at least three different sorting functions for the episodes in the saga (release order, episode order, and machete order[1]), each unfolding a different story line but all still *Star Wars*.

Finally, a few words on learning. I liken innovating to learning—the learning that occurs while you are engaged in a very general kind of problem solving, with no guarantee that you will come up with a solution. This is extraordinarily liberating. The operating question isn't "How do I apply this framework to that?" but rather "What is the knowledge or evidence I need to acquire to make that problem real?"

Innovating by making real-world problems tangible offers you an alternative to the many innovation recipes that have emerged from product design, product marketing, lean manufacturing, and technology readiness—the recipes my students have in mind when they ask the "contaminated" questions. To get started, all those recipes seem to require a well-formed idea about a product, a user base, or an organization—that is, they require that a large part of your innovating be fixed before you can even begin. That feels suffocating to me. The urgency to "productize" every observation feels unnecessarily constraining, and the rush to drive every action toward identifying product placement opportunities feels like opportunism. Most of these recipes seem to take a "good idea" as a given, and hinge on convincing others that it is, indeed, good—and then placing it.

That is quite the opposite of learning. This book is about learning.

My training does not seem to have prepared me well to produce new "good ideas" from the get-go. Instead, it has prepared me to arrive at them. I don't particularly care for processes that put me in the somewhat weak position of having to convince others that an idea is good before I myself am persuaded. And I am not particularly motivated to perform tasks that are presented primarily in terms of pleasing users or designing to their liking. I would rather solve a real-world problem.

What I, and most of my students, seem to get are intuitions about problems—*hunches*—and a desire to learn why our first intuitions are wrong. Some students feel compelled to present their hunches as if they were products; but they are just that, hunches.

Whether you ultimately decide to engage in innovating by focusing on making a problem tangible, as I propose and explain in this book, or instead to follow one of the innovation recipes is your choice. This book doesn't replace those recipes, which serve a purpose. The approach presented here is, though, wholly different; the concepts undergirding this book borrow nothing from those other recipes. Working on the problem requires a new vocabulary, a different attitude toward innovation, and a subtle mind-shift.

The book is a manifesto for doers to embrace their doing as an instrument for exploration. It is also an explorer's guide into the impossible; in fact, I see innovators as the explorers of our time. The book shows a path for exploration: Accept that you will learn by being wrong as you venture into the impossible in search for that thing others will come to appreciate as magical—an "innovation"—when your turn comes to tell your story in hindsight.

To be clear, I am not taking a position on whether innovation can or cannot be learned. I just think it's a moot discussion. I hope to persuade you that innovating, like most other activities, is something you can practice and become better at with the right combination of knowledge and the kind of muscle memory that comes from repeating certain tasks—that is, from *doing*. Innovating takes doing, practice, and perseverance—which are how your brain has adapted to learn best. At some point, you ought to learn to trust that your brain can operate quite well outside the realm of formulas.

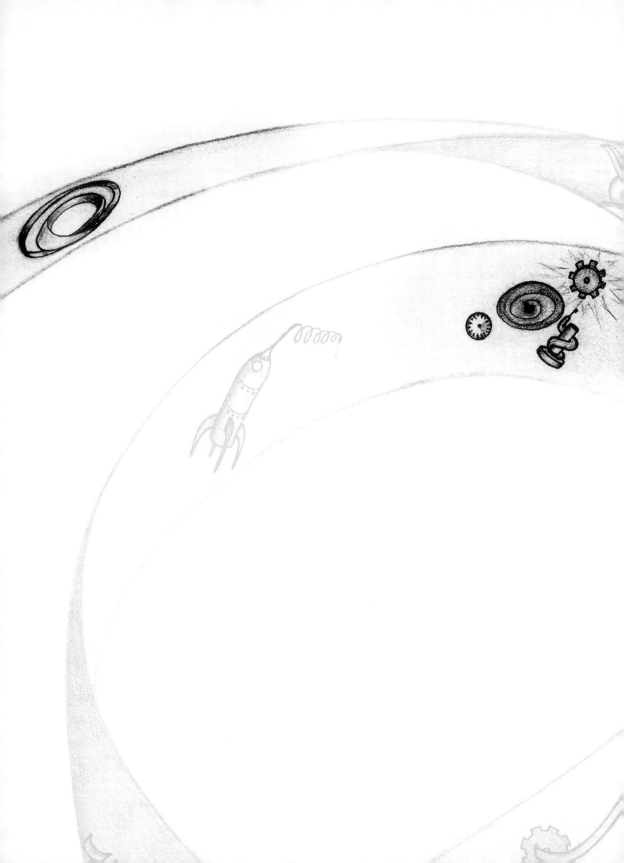

I

ANATOMY OF A HUNCH

PERHAPS YOU HAVE AN INTUITION about an "innovation." Today, it may look to you like a solution—defined by a product, a cause to champion, a technology, a user need, a process, an environment. It may look impossible. It can become an intuition about a problem. For now, it is just a hunch.

Eventually you may arrive at an innovation. But there is more to innovating than the hope to land on *the* innovation. Your forecast is that there is an opportunity to do something that has not yet been done, that a real-world problem has yet to be solved. At the outset, you begin with nothing more than a hunch and what you already have. You can innovate from there.

You may be inclined to bank it all on the belief that the *one* solution you imagine is right. But you do not need to.

Your hunch also hints at a broader space of opportunity. Most likely, there are multiple ways to have an impact in that space. The solution you imagine *may* be one, but for now that's at best an assumption at risk of becoming a significant constraint.

At first, it may be more straightforward to find all the ways your hunch cannot possibly work than to try to land on an innovation deductively. You can learn all the ways you can solve the problem by allowing yourself to be wrong about your hunch and finding out how you are wrong. The outcome can be a robust path to solving a real-world problem—no matter how wrong you are at the outset.

Whether your innovating leads to *the* innovation you imagined, you stand to benefit the most from going about it with a certain naiveté.

You can become increasingly better at *innovating* through practice. All it takes is learning to distinguish what you ought to do from what you'll eventually produce. The chapters in this part of the book introduce principles for doing so. In a sense, these chapters are the entire book; the chapters beyond expand these principles into that practice.

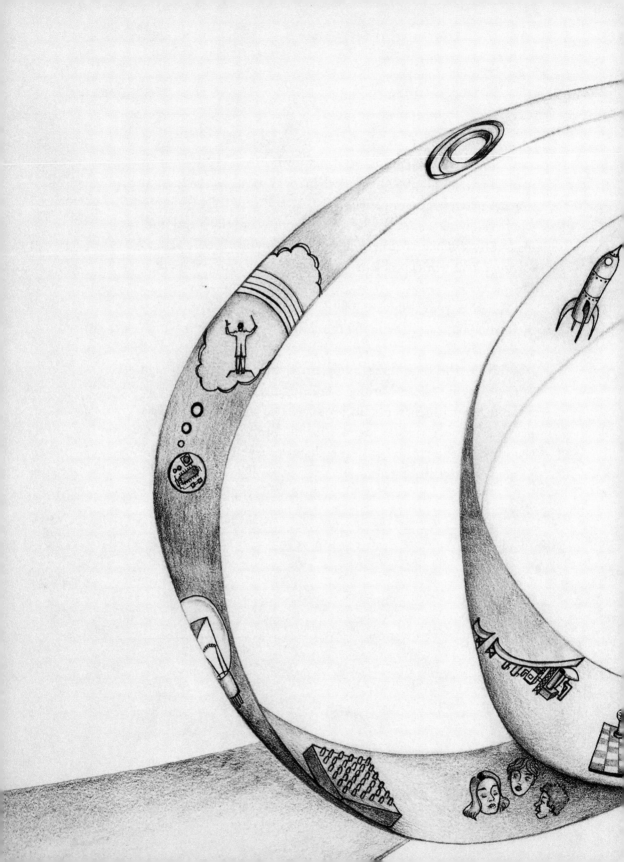

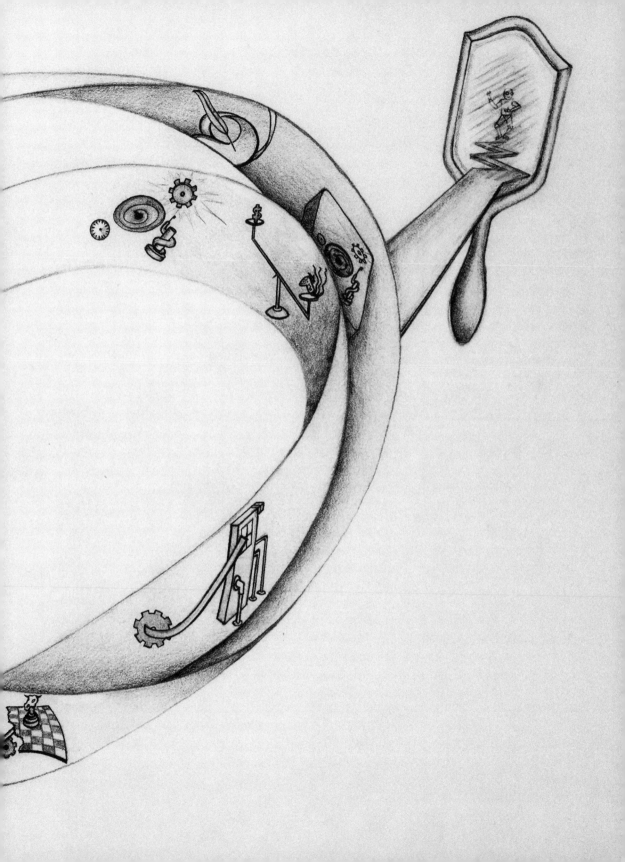

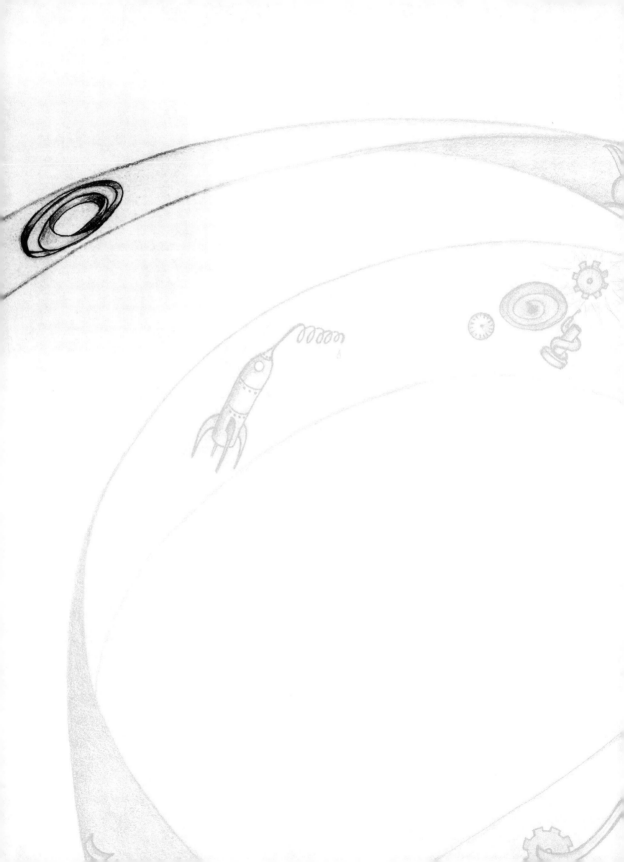

BEING PRODUCTIVELY WRONG

At its genesis, no thing about an innovation is new.

This observation will likely seem flat wrong to you. It did to me. Over and over, I dismissed it as a curious if not irritating paradox, until I finally decided to look at it in a different way. Then it became a relief: that novelty will reveal itself. Let me try to put that another way: Innovations accrue their novelty as you innovate. They are more easily deemed innovations in hindsight than at their beginnings. In hindsight they can be judged by how they ultimately empower others—a community—to achieve new things.

If you accept that innovations hardly look new when you begin, you can stop searching for the next earth-shattering idea and instead just start looking around for a way to fix that thing you already suspect is not quite right. As you do, you might unknowingly start on a path to change the world.

Still, something about an innovation not being new at the start feels paradoxical. I have come to find solace in time travel.

Rewind. You are now in 1960. You've recently seen a great comedy, *Some Like It Hot*, at the movies. People are still debating the "risqué" scene in which a man dressed as a woman kisses a woman. You missed it because of the high hairdo of the patron in front of you, but you're still out 70 cents for the ticket.

July 8, 1960. You pick up the *New York Times* and scan the cover. You read about John F. Kennedy advancing in two states ahead of the primaries; rivals dispute the claims of the "New Englander's aides." The story continues beneath the fold. Underneath it, in the lower left corner, a headline reads "Light Amplification Claimed by Scientist." Next to it, near the bottom of the page, a big photograph of Congolese troops and demonstrators and a story about fleeing Belgians jumps out at you.

You sigh. News as usual—politics and war. You put down the newspaper and decide to listen to a phonograph record.

That article about the scientist, easily missed, reports on an experiment from May 16: physicist Theodore Maiman at the Hughes Research Laboratory focused a high-powered flash lamp on a ruby rod. It is the first device ever to emit coherent light. Dr. Maiman just demonstrated the first working L.A.S.E.R. (light amplification by stimulated emission of radiation). You are still in 1960; the significance of this news is, at best, elusive.

Fast forward. May 6, 1964. On page 69 of the *Times*, you catch an interview with Dr. Maiman, now president of Korad, a subsidiary of Union Carbide that builds lasers. He describes the laser as "a solution seeking a problem."

Now it's the 1990s and you're on your way back to the present. You see a TV interview with Dr. Maiman. He's holding that first laser and reminiscing about his feat:

> One of the things I at least attempted to do and it did turn out to be successful was to use things that were already around. ... If I had to develop a special lamp that would be a whole little research project there. ... I had a crystal that's not only found in nature but can be made very pure industrially. So, I was able get a hold of some rubies and just order them, buy them and have them cut. The lamp I could buy [from a photographic equipment catalogue] and the rest of it is just simple machining.[1]

Back to the present. The laser is so pervasive you may not even know your daily routine depends heavily on its existence.

It is disorienting that no part of that first laser was new. The laser itself, though, was. Theodore Maiman created his "invention" using parts he either found around the lab or could order easily from a supplier. The laser was still a long way from being recognizable by its name the way it is today. But it was an instant innovation within several close-knit communities. Within weeks, several research groups were using variants of Maiman's design. Shortly after his discovery, Maiman and some colleagues left the Hughes Research Laboratory for a company that produced ruby crystals, then founded Korad to sell

At its genesis, no thing about an innovation is new.

lasers. Within a few years, lasers began to be used in welding, medical research, and so on. One of the first applications reported for the laser outside of research in optics appeared in 1961 in a clinical research paper assessing its use as a replacement for a scalpel in eye surgery.

Nothing about the first laser was new—that is, not a single part, not even the first domain of application.

Might innovating really be that simple? You and your peers put some parts together? Someone else, unconcerned with your parts and process, adopts your artifact and derives a benefit from its use? There you go—you just produced an innovation. And all you did was iteratively assemble a few parts—some technical and some not, initially none new—in a new way, and drive them to adoption by someone who is an expert at something else.

It still took nearly two decades for the laser to find its way into a mass-market innovation like the ones we admire and have eventually come to take for granted—specifically, as a component in bar-code readers. Along the way, lasers made it into innovations in all sorts of contexts, in academia, and in numerous other industries. In 1962, a laser pulse traveled to the moon and back for the first time. In 1969, the Apollo 11 mission installed the first retro-reflector on the moon, and lasers have been used to measure the distance from Earth to the moon ever since.

But that's the laser story from the beginning. There is another way to tell that story, and it is the way "innovations" are often described: working back from the *present*. From that perspective, the laser looks quite different. It is no longer a note about a scientist in the *New York Times*, or a "light amplification" device produced by a researcher with mostly "old" parts and nearly no budget, or a technology that went from cover to page 69 in four years, all the while becoming a "solution seeking a problem." It is *the* laser—a technology that revolutionized several industries and society itself, and one that we now take for granted. Looking back from the present, it is easy to imagine the device being interesting from the outset, to fictionalize in our heads a dramatic competition to be the first to make it happen, and to imagine many people working on it. It is a gripping tale.

In fact, that story is also true. It is the story of how our society first encountered lasers. Billions of dollars were indeed spent on research and development of the laser before and after Dr. Maiman's first laser. Jeff Hecht's book *Beam*[2] thoroughly documents the history of the laser as a societal development emerging from a community of scientists—a "race to make the laser." But in that version of the laser story, the specifics of how the first laser actually emerged become somewhat anecdotal, as though the laser was bound to happen. That invites a question: How significant is it that the first laser was built out from readily accessible parts, seemingly as a scientist's side project?

Most of the innovation stories that inspire us share one fundamental characteristic: What makes them exciting is that we already know the end. They make it easy to imagine the first steps: "get an idea" and "assemble a team." Those are steep first steps. I contend, though, that the very notion that at the beginning you can even *identify* an innovation is specious at best. That's why I have come to appreciate that the *seemingly anecdotal* story of how the laser actually emerged is so significant.[3]

As you embark on innovating, the choice of which version of the laser story to draw from for inspiration should be your own. Both versions derive their meaning from the same laser. You may start by putting together a few parts to make sense of a problem that is apparent to the members of your community and perhaps to no one else, at least yet. Or you may try to identify *the* innovation that is *bound to happen.* I find the hindsight version of the story stressful, nearly paralyzing; it seems to imply that the kickoff is an *earth-shattering idea*, one you get with virtually nothing to go on but your imagination. Powerful as your imagination is, though, it does not hurt to aid it by giving your hands something else to do besides holding your chin. I prefer to draw inspiration from the story that looks forward, in which being wrong is as natural as getting things right.

The forward-looking story of the laser suggests a strategy for innovating that requires only moderate foresight—a hunch—and gives you something to do right away: Just put *some* parts together. With the wealth of resources available online today, two-day shipping, the availability of MakerSpaces, and the increasing access to moderate amounts of seed funding, that approach is more achievable today than it was in Dr. Maiman's day.

Imagine yourself in the 1960s. You and some friends are sitting around discussing whether it could ever be possible for a group of civilians to stop the world's most powerful government from testing the world's most powerful weapon of war. A few years later, that is how Greenpeace will be born, after challenging—with a fishing boat—the US government's plans to test nuclear weapons.

In the years following 1969, more than thirty people came together in various forms for some time, driven by a desire to make a difference about the environment. They converged at different times in several groups driven by

> What makes most
> of the innovation stories
> inspiring is that we
> already know the end.

ecological consciousness: the Scientific Pollution and Environmental Control Society (SPEC) in Vancouver, British Columbia; the Ecology Action Group in Berkeley, California; the "Green Panthers" in Vancouver; the Don't Make a Wave Committee in Vancouver; and the Sierra Club in Vancouver and California. This loosely bound collective was trying to figure out what to do to bring ecology and the environment to the forefront of society, creatively, through action.

At first, participants borrowed from the playbook of other ecology groups in California and Vancouver. They staged creative protests, wrote columns in the press, and put up billboards in Vancouver, guided by the thought "If you can promote companies and products, you can promote ideas."[4] The group even bought an old boat for ecology actions. It eventually sank.

In the summer of 1969, the US government announced a plan to conduct nuclear bomb tests on Amchitka, a volcanic island in the Aleutians off southwest Alaska. This announcement gave several participants a definite purpose. The group, led at different times by Bob and Zoe Hunter, Irving and Dorothy Stowe, and Ben and Dorothy Metcalfe, set off to stop a nuclear test scheduled for the autumn of 1971. It would be the second such test.

In 1971, members of the group chartered a small fishing boat and set sail for Amchitka from Vancouver. Proceeds from a concert helped fund the trip, and several established ecology organizations supported the trip in other ways. During the trip, those on the boat communicated with group members who stayed behind via a radio at one member's home. The stay-at-home members served as a media link. Two weeks after departing, well before reaching the blast zone, the boat was intercepted by the US Coast Guard. A month later, the test went off as planned. They failed to stop it; they never even made it to the Amchitka blast zone.

The failed journey, though, garnered an unexpected amount of news coverage and support from society and politicians.

That adventure began with a hunch about a social problem, billboards, membership in environmental organizations, protests, op-ed columns, a concert, a fishing boat, a trip to Amchitka, a radio, and a link to the media. It succeeded in bringing ecology and the environment to the forefront. The trip to Amchitka brought together these "parts" in a new way and persuaded several of the participants that there was a sustainable path forward for an organization concerned with a new way of engaging with ecology. Months after the failed journey, they renamed the Don't Make a Wave Committee the Greenpeace Foundation.

Just as with Maiman's laser, nothing in that first trip to Amchitka was new. The combination of mostly old parts yielded something many recognized as new and impactful. It was, in a way, a fully working innovation prototype. It demonstrated a new way of engaging in ecology through action, it resonated with a much broader community, and it emerged from an embryo of an organization—a collage of organizations, actually—concerned with both the trip and the diffusion of its meaning. But it was also, in a way, just the latest and most widely successful action of an evolving group of people engaged in what might be described as a sequence of trial and error to find a sustainable way to equate ecology and action. They returned absolutely convinced they had failed. They were wrong about their impact. The unexpected response to their actions showed that it was not necessary to stop the US government to make a difference. It could have stopped at that, except several members of the group were persuaded that the trip to Amchitka was, in fact, a scalable recipe for ecology in action.

Their next iteration did not affect the recipe for the trip. Rather, the experience they accrued from participating in several ecology groups and the trip to Amchitka informed a new organization—the Greenpeace Foundation. Its founding, however, did not put an end to learning through trial and error. A year after the first trip, the Greenpeace III sailed toward the Mururoa Atoll to stop French atomic tests. This first action of the Greenpeace Foundation was, fundamentally, a repeat of the trip to Amchitka. The founders of Greenpeace reminisced in a 1996 interview[5] that the problems surrounding the trip of the Greenpeace III taught their nascent organization a lot about the legal implications of this new form of ecology action, its international ramifications, and how to recruit new activists. In essence, they were learning how to bring the new thing called Greenpeace to "scale."

Only in hindsight does it seem as if the founders of Greenpeace had come across one of those earth-shattering ideas. Looking forward from the beginning, though, it seems their clarity about their group's purpose—"ecology through action"—was not a given; rather, it emerged progressively through trial and error, from actions before and after the founding of the organization Greenpeace. In fact, the specific problem Greenpeace describes as its mission has evolved continuously over the years. And several of the organizations that shared a similar goal around the time of the founding of Greenpeace—and at times also shared the same people—never scaled up the way Greenpeace did.

The laser is generally framed as a technological innovation that first had a profound impact in research. Greenpeace's ecology in action is framed as a social innovation. Their outcomes certainly are technological and social. Their beginnings imply there may be something more general about how innovations emerge. We can pretend-play to time travel, but again we can see the laser and Greenpeace as innovations only in hindsight. Their stories provide little in the way of specific lessons about how to innovate successfully. The only things we can say for sure about what they teach is that you need a hunch about a solution to a problem, some parts and some people, and, ultimately, a willingness to be wrong as you try to accomplish something. The "be wrong" part is the way to discover which parts to keep, which to discard, and which to acquire that you don't yet have, along with which people you need and which people you don't as you evolve your hunch.

Getting started innovating may require abandoning some preconceptions about how you think about innovations once they reach society or the market and instead thinking of innovating as something you start to do way before the idea of a product or an organization eventually reveals itself. With that in mind, let me tell you another brief story, again from its beginning—in November 2007.

It is November of 2007. A graduate student in computer science posts a video on YouTube in which he explains how to modify the remote control of a video console using an infrared sensor so that it can detect the position of the user's fingers. The remote is held in place by a deck of cards. The video refers to a website where he makes the code available for free. On the same website, he also shares instructions on how to use the remote control of the console to develop an interactive whiteboard and track the user's head to develop virtual reality simulations. This is a side project, a "distraction" from the student's thesis work. There are parts, and there is a virtual community—the YouTube audience. YouTube was then about two years old.

Think of innovating as something you start to do way before the idea of a product or an organization eventually reveals itself.

Now here is the hindsight version of the story. In 2011, Google hired a computer scientist by the name of Johnny Chung Lee who specialized in human–computer interaction and was best known for his work to extend the functionality of the remote controller for the Nintendo Wii video-game console. He's the same computer scientist mentioned in the preceding paragraph. In between posting the video and his employment at Google, he worked at Microsoft, joining as a core team member of the project that came to be known as Kinect. You can get the full "story," backwards, at any number of websites.

In the years since Johnny Chung Lee's innovation, the Kinect has come to be used in personal fitness, video games for dancing, and a number of other action-based video games. Video games and fitness, once not even remotely connected, are today intertwined. That, though, is something you can only know after the fact. Still, it doesn't stop people from pointing to the Kinect story as an example of a "truism" about innovation: *focus on the user*. Of course, it's a lot easier to do that when you have a clear image of the end product than when you are making an early YouTube video. In the former case, the user is obvious; in the latter, the user is fictional—at best, another enthusiast. But that didn't prevent Johnny Chung Lee from diffusing his innovation and being recruited into a Microsoft team.

The Forward-Looking Perspective

If you truly adopt the forward-looking perspective, certain things that might otherwise seem easy to do reveal themselves as quite cumbersome. For instance, "focus on the user" is a rather abstract instruction. Finding users implies they exist. Calling them "users" implies they are using something, which obviously can't be your product if you're at the beginning of the innovating process. Your users don't exist.

The paradox of focusing on or finding the user *at the beginning of the innovating process* is that while someone may *eventually* use or benefit from what you have to offer, there is no way you can design a yet-to-exist product around non-existing people. To do that, you must have set or fixed some users and some kind of product. The level of certainty around products and users required to undo the paradox and allow you to engage in genuine user-centered design is fairly high. Before engaging in such design is even possible, you have two choices. You can arbitrarily restrict your attention to an imagined product or you can engage in a significant process of discovery that will reveal the offering and the beneficiaries, along with many other things. I advocate the latter.

Before you have real, not imagined users, your innovation will look like a bunch of old parts strung together. If you're like me, that's a relief.

The Kinect story teaches the same lesson as the laser and Greenpeace stories: you need some parts, some people, and a willingness to keep trying or "tinkering" while you learn from being wrong. Depending on when you stop the reel, the innovation is a Nintendo Wii "hack" in a YouTube video, and the outcome is a series of TED talks and employment at Microsoft, or the innovation is the Microsoft Kinect, the team is the group Johnny Chung Lee joined inside Microsoft, and the outcome is myriad new opportunities for video games.

That's enough time travel.

I have walked you through three examples involving technology, social purpose, and a product to persuade you that the beginnings of an innovation are a mismatch for the stories of the organizations and innovations they ultimately empower. All three stories have beginnings that seem to have more in common with stories of failed attempts at doing something than with the companies and organizations that grew out of them. In fact, had I concealed the details of the eventual innovations any more than I did by playing time travel, you may have thought this book was little more than a historical compendium of people engaged in a kind of hacking for no apparent reason. That's the beauty of innovation: At first every project looks like every other project, no magic or foresight is required at the start, all are welcome, and then the final result bestows recognition as an "expert" and makes the story inspiring.

The stories teach that the path to impact—whether the outcome of your innovating is a product, a technology, or a new kind of organization—appears to be fraught with near misses and learning. But at the outset, no thing is new. There isn't yet a product or even a concept of one, the minimum viable "product" is "do nothing," any notion of a team is at best fluid and at worst just a loosely bound collective, users are fictional characters, the organization has yet to be determined, and the idea is not yet earth-shattering.

That's good news. It means that right now you have everything you need to get started, and probably more.

It also means that you need an approach to learn as you go until the product, the organization, and the problem you solve have become clear enough and serving your solution at the right scale has become your main worry.

The reason all this matters is that if you accept overturning some of the conventional wisdom about innovations, as I've tried to do, you can begin innovating without worrying about all the considerations that matter only in hindsight.

Learning

A "minimum viable product" is also a highly abstract concept. Viability is judged at the receiving end. A product is a highly elaborate construct. And let's admit it: what is minimally viable depends largely on your audience's tolerance for lack of quality.

A more helpful forward-looking perspective is that no matter how your first offering or prototype looks, absent room for improvement there is nothing left to learn. The outcome of each iteration is the next reduction to practice of your idea. As you work your way through iterations and uncertainty, each successive prototype gets you closer to being able to apply principles from management and design. Sacrificing quality to get something out sooner is not a principle you should adopt. Learning to increase quality with each iteration is.

On top of that, the moment your product requires capital expenditures, what had at first appeared to be minimally viable may actually require as much investment of labor and capital as the next higher-quality alternative.

Right around the time I introduce these ideas, my audiences begin to push back hard and start asking me to define what I mean by innovation. I suspect most readers would do the same. The most adversarial members of my audiences insist "That's not an innovation, because not everything is new." Others want to know how much "new" is needed in order for something to be an innovation.

I actually think defining innovation is a distraction from innovating. But I typically give in and offer a simple relation as a definition, which should suffice for our purposes here.

innovation: novelty with impact

But this is the "definition" of what you end up with. You evolve toward it; it is not a given, and for that matter it is also a rather bad indicator with which to measure progress. At some point down the road, you'll bring something new to a community and its members will benefit from it. Their benefit will be your impact.

It is the final result that turns someone into an "expert" and makes the story inspiring.

I have intentionally left out of this relation both technology and the market—or technology push and market pull—two of the usual contenders for innovation. Depending on the scope of your innovation, they may become really important. At the beginning, though, they just get in the way. You choose the technology that will empower the novelty, the impact, or the interaction between them; you choose the market in much the same way. The magnitude of your impact is a measure that depends on the community you intend to serve and the scale at which you bring the benefits of your innovating. All these are, as a matter of fact, consequences of the choices—both conscious and unconscious—you make as you continue learning how to turn your hunch into a solution to the real-world problem that gives you purpose. If you set or fix any of market, technology, product, or impact early on, with nothing to go on but a hunch, the nature of the task at hand will change: you'll soon find yourself unburying the assumptions you unknowingly made.

> The path to impact
> appears to be fraught
> with near misses and
> learning.

Let me give you a few quick examples to illustrate my point about technology, markets, scale, and impact as choices.

The first example concerns Henry Ford, who early in the twentieth century helped lift much of rural America out of poverty with his affordable car. His innovation required conceiving a new way to assemble cars—the assembly line—and new approaches to empowering buyers so they could obtain those cars, including financing their purchase and raising his worker's wages. At the outset, however, Ford likely concerned himself with a more specific problem: the relative isolation of rural life and the time it took to go to the nearest town for supplies, which he knew from his own experience growing up on a farm. That problem would have already been solved by cars, which already existed, had they been affordable. It is unlikely Ford set out to invent the assembly line, introduce auto financing, or raise wages. Rather, those were consequences of the choices he made.

Henry Ford's "Hunch"

Early in the twentieth century, Henry Ford's venture lifted rural America in ways we would like to reproduce today worldwide. His feat is often associated with technology and the assembly line and with innovation in management and financing, and arguably less often with its profound social impact. But at the time, cars were not new, raising wages five-fold was against the conventional business wisdom of the time (as it is today), and the technology Ford developed was really an instrument to scale up an idea: an affordable car to empower rural America. By making cars affordable to workers and farmers in small communities, and by financing their purchase, Ford did as much to change society as he did to change the future of manufacturing. Ford's story of social change has unexplored parallelisms with Grameen Bank's empowerment of the poor at the turn of the 21st century through microloans.

But Ford's story really begins when he was a self-taught employee of the Edison Illuminating Company and built a quadricycle (a "hacked" car) in his shed to persuade himself first that the affordable car he envisioned could be built. His finished quadricycle did not fit through the shed's door, so Ford had to smash the doorframe, and then the quadricycle came to a complete halt a few blocks down the road. Consistent with the trial-and-error nature of innovating, several iterations later the Ford Model A (1903–04) reportedly suffered from overheating and slipping transmission bands. Eventually, Ford got to Model *T*.

At the beginning, all Ford had to go on was a "hunch" that if cars were more broadly available, the problems he had experienced growing up on a farm would disappear.*

* For a historical account of the many iterations in organization, management, and car models Ford went through to refine his hunch, see the episode of the PBS television series *American Experience* titled "American Experience: Henry Ford" (available at http://video.pbs.org/video/2329934360).

Next up is Greenpeace, not usually thought of as a technology organization. However, as Greenpeace learned how to organize to serve its environmental purpose, those involved made use of several technologies that had recently become available for such use, such as long-distance radio communication between a ship and someone's home.

Finally, there's the Polymerase Chain Reaction (PCR), a technology developed in the 1980s. It was instrumental in the rebirth of genetics as a field at the crossroads of biology and information. In the 20 years after its discovery, the impact of PCR materialized as a Nobel Prize, new instrumentation, and the Human Genome Project. Each was a consequence of decisions made by certain individuals. Whether PCR was itself an innovation and when it became so seems subjective.

When Did It Become an "Innovation"?

The Polymerase Chain Reaction (PCR) is a technology invented in the early 1980s. Today, with the benefit of hindsight, you might describe PCR as a reasonably simple process for amplifying a genetic signal—but you can do that only because we now know what to use it for. In fact, that description identifies genetics as an information science, something it was not considered to be when PCR first appeared. In its early days, PCR was described only as a way to massively duplicate molecules of DNA.

PCR led to a paper, a patent, the sale of a patent, and a Nobel Prize, and to the development and commercialization of new instruments (e.g., thermal cycling machines, openPCR, and PCR chips). It enabled the modern field of genetics and the Human Genome Project, Ultimately, PCR's very existence as a tool spawned new research directions, created a need to develop new educational curricula, and made it increasingly possible to conceive of genetics as an information science. All these outcomes combined technology and management considerations to different degrees.

So when did PCR become an innovation? That is a question I ask in my lectures to make a very straightforward point, one often overlooked. In hindsight, it is customary to conflate the moment of invention with the moment something "became an innovation"—as if it had been an innovation all along. But when I ask the question about PCR, there is disagreement in the audience about which event "turned" PCR into an innovation. To some it was the paper, to some the Nobel Prize, to some the development of a commercial use, and to others the emergence of the thermal cycling machine a product.

PCR demonstrates that innovation seems to be in the eye of the beholder. What you do next with your hunch depends largely on the community you choose—which, incidentally, means that the vehicle to scale up your hunch into an innovation is also a choice.

In a 1984 interview, Steve Wozniak, a founder of Apple, reminisced:

Those of us involved with micros back in 1974, 1975—we had not already designed all ... that goes into the mini computers of the day or the largest systems. To us, it was so exciting because ... we thought we were doing it for the first time it had ever been done. All it was, it was the first time they had ever been done that cheap ... showing off a little card that would play music on a computer or make color ... we thought we were way ahead of the rest of the world. ... I was not even thinking about what are the right steps to take to have a very large successful company or a large successful product. ... I had been working my whole life to build a certain type of computer for myself, and I just built the best one that was doable in that day with the particular components available.[6]

Indeed, Wozniak first shared the computer schematics for the Apple I in a small interest community known as "the homebrew computer club"—for free.

The stories I have shared with you thus far span innovation across technology, market, management, organization, social purpose, and product. Their end points have in common only that they all had lasting effects on society. Their beginnings seem to share that someone had a hunch about a problem. The hunches are specific, but the problems "ecology with action," "uses of the laser," "user interface like in the movie *Minority Report*," and "affordable cars" would allow for many solutions. As Dr. Maiman put it for the laser, at some point all of those hunches look like "ideas seeking a problem," not yet like *innovations*. What seems to turn these into innovations is the focus on the particular hunch and the determination to see the problem through. The sense of a larger opportunity in the making seems to emerge as decisions are made and reversed. The honorary recognition as an innovation arrives much later. At the beginning, only the hunch contains some essence of novelty.

It is easy to get caught up with the final result we imagine—the *innovation*—and believe it ought to be recognizable as such during the process. I call it the curse of innovation because it distracts you from everything you need to learn to drive your hunch to become a solution to a problem—it distracts you from *innovating*. Instead, it urges you to rush to productize your fragile hunch when, in fact, what you might need to do is accept that while your hunch may be correct, your first idea of how to materialize it is most likely wrong—just as you want it to be.

The world looks tremendously different from the perspective of a finished project than from the perspective of a budding project. There are multiple paths to getting to impact, and multiple forms of impact. None are known until your turn comes to write the story of your innovation in hindsight.

The examples I have shared—Laser, Greenpeace, Ford's assembly line, PCR, Kinect—help make my points about innovation. Even when the outcome is well known, selecting the technology and predicting its readiness both seem quite difficult. The Ford example teaches that innovating is about far more than technology selection and management. Absent background information on what happened, predicting the outcome from the input is quite difficult, as the Laser, Greenpeace, and Kinect stories suggest. The PCR example shows that recognizing something as an innovation truly is in the eye of the beholder. More than that, it's generally a revisionist exercise: Again, we can only define an innovation after the fact.

The PCR example also shows that the vehicle to impact is a choice. That choice further implies the choice of a community. For Apple, it was a startup. For the laser, it was first academia and then it evolved rapidly into a startup. PCR first became a patent sale. Ford and Greenpeace—one of which became a successful for-profit company, and the other a non-profit organization—both began by tackling a social problem. The hacked Wii led to employment at Microsoft and to the Kinect.

None of these examples suggests a linear process in which changes at the outset can easily be traced all the way to the outcome, much as you expect a car to accelerate mildly after a gentle push on the gas pedal. We humans seem to strive to find linearity even when there is none. That's why stories in hindsight that paint a linear path are so soothing—no matter their seemingly magical changes of direction.

Innovating, by contrast, is a highly *nonlinear* process. Apparently tiny changes to the input can translate into wildly different outcomes, suggesting to an outside observer that you somehow effected a magical change of direction. And yet that's precisely what we aim for when innovating: presenting society with a seemingly small change that will change the community that receives it in a profound and lasting way.

Your surroundings are full of other examples of nonlinear processes. They all feel somewhat counterintuitive. Suppose, for example, that your next gentle push on your car's gas pedal returns a loud roar from the engine and sends the tachometer toward the red line. Having expected a linear response, you would likely think it is time to take the car to a mechanic.

In the case of innovating, there are some noticeable advantages to this nonlinearity. Early on, with only small changes to what you are building, you can explore wildly different kinds of impact. Conversely, forms of impact that look similar at the outset may in turn be enabled by very different solutions. That is, early trials can reveal a lot of information about the problem you set out to solve.

But if you are hoping for a straight path to impact, innovating may appear daunting at first. You need a lot of information to trace changes at the outcome all the way back to the beginnings. That's why the stories of innovations in hindsight reveal so little of what one needs to do. And forecasting an outcome, or a product, or a user, or an organization, or a business model, or

> None of the multiple paths to impact are known until your turn comes to write the story of your innovation in hindsight.

the specific technology needed from the hunch that characterizes the genesis of an innovation requires obtaining an insurmountable amount of knowledge of the dynamics ahead.

At the beginning, all you have is uncertainty. You do not yet know what, if anything, has to be true. And that uncertainty—whether it's conscious or not—is probably why you are embarking on this adventure. What you must do is discover what has to be true about every aspect of the problem surrounding your hunch. All this invites a question: If things are bound to change so much, how much of what you do next should be based on the endpoint you currently imagine? My short answer is "Nothing." Otherwise, you will face one paradox after another and you'll end up overworking.

Paradoxes

At the beginning of your journey, several concepts that will become paramount as you get closer to an actual "innovation" should lead you to paradoxes. The literature on entrepreneurship and innovation is rich in insight into specific attributes that have defined success for many ventures. "Focus on the user," "identify your lead users," "your job is to create and capture value," "develop a minimum viable product," and "experiment," to name a few, are shorthand we use to refer to a much richer reality. For the most part these statements are true, but they are too influenced by the end of the story to be as useful at the beginning as they might be when you get closer to an actual "innovation." They are, for the most part, statements about desirable outcomes, and are largely things that will *eventually* come true (they describe what you will someday have), but not one of them informs you about what you actually need to do now to make them come true. And that is why you'll end up overworking if you accept them.

With the knowledge available at the beginning of your journey, the statements above should look like circular references, as in "focus on the people that will use your yet-to-exist product." Even purported actions such as "pivot," "iterate," and "select the best ideas" really just describe what happens rather than what you ought to do and what you ought to learn. They fit very naturally with the story from the hindsight perspective, but they offer little.

The same goes for "get a team and an idea." You can buy parts, but "ideas" don't come "pre"-labeled as worth exploring and people don't come "pre"-labeled as "team."

"It's all about management" raises the question "Managing *what*?"

"Focus on disruptive innovations" seems like an admonition to travel forward in time, discover what will disrupt, and then return to your own time to make the big score.

Words such as "user," "need," "business," "value," "innovation," "team," "pivot," "iteration," and so on acquire relevance and specific meaning for you only through the evolution of your idea. Their meaning may appear to have been fully crystallized when you inspect a company from the endpoint, but that is an illusion. They lacked any real meaning at the outset. Again, those words do not describe directly what you do. Rather, like in the stories in hindsight, they are part of the summary of what happens. To apply them, you will have to set in stone many of the attributes of your hunch with little evidence to go on. There is nothing inherently bad about that, provided you stay honest about what you set in stone and allow yourself to revisit those choices you made without evidence.

What you *do* need to innovate, however, need not be complicated. That's not what nonlinearity means. Nonlinearity simply requires you to accept that early on, your forecasted end-point is almost certainly wrong. You need to expect to be wrong. When something doesn't work, you know it. That knowledge is infallible. When innovating, you learn by being wrong.

Innovating as learning

The process of creating what eventually becomes an innovation is something you can learn and become better at through practice. In fact, if you can overcome your concern at the beginning for whether an innovation actually results, innovating and learning can be the same thing. You learn about the problem you are solving as you innovate.

The kind of learning I'm encouraging you to embrace is a kind that was more natural to you in your childhood. You are allowed to use your intuition, engage in trial and error, and be productively wrong. That's how, in fact, you acquire language. Grammar comes later. As you grow older, instruction becomes more formulaic. How does this play out with innovation? A "formulaic innovation learner" wants to know what kind of idea he or she should have and wants a recipe to turn that idea into an innovation. That implies an idea already exists and that there's a team to "execute on" the idea.

The fatal flaw in such thinking may not be obvious, but it should be. It is that the idea must be largely fixed if the recipe is to work. The recipe for a chocolate cake works only if the idea is to bake a chocolate cake. Following that recipe will not result in lasagna. The recipe for the trip to stop the nuclear test in Amchitka does not immediately lead to an organization. Indeed, the *Golden Rule* expedition tried to stop a nuclear test ten years earlier in much the same way—and no Greenpeace emerged.

With the illusion of innovation powered by a fixed idea and an *a priori* recipe, the discussion quickly turns to things you're not ready to decide— curvatures, size, whether target users need deep pockets, and so on. These are all important considerations in certain contexts, but the idea had better be right, because none of these will fundamentally challenge it. Your innovation is going to be a car, or some kind of chocolate cake, no matter what. But if you are still open to your innovation being social, industrial, system, design, manufacturing, regulatory, academic, educational, organizational, or technological, to name a few outcomes, then the process you follow ought to allow you to discover that as well.

Then there's the obsession to be "disruptive" with an innovation. Fixing the idea places the burden to be disruptive on the idea itself. This constrains your options. Putting a thin layer of metal inside a light bulb launched the entire field of electronics. That led to the vacuum tube, the impact of which was disruptive. The initial technology was incremental.

The good news is that there's a way around the madness. This book is about what I call *innovation prototyping*, an incremental, nonlinear, and highly experimental process predicated on evolving solutions to real-world problems together with the means to bring them to the point where they have an impact. That impact might be disruptive, but what you do through your innovating does not have to be. What you do can be stated quite simply: Summon up a hunch about a problem, gather parts and a group (a "community") to work with the parts, and then work to have your solution scale. It all starts with a desire to demonstrate tangibly first to yourself and your community that the problem you are targeting can indeed be addressed.

The entire process works the same way you acquire language. You learn to speak by talking. You learn to innovate by innovating. You're allowed to use your intuition, engage in trial and error, and be productively wrong. The process does not start with an idea for an innovation. It begins with something that is closer to you, something you cannot explain, something that needs fixing, or even a pet peeve. That is the problem that gives you purpose. Your job is then to make the problem tangible. You do that by questioning the problem continually. The parts and people you gather through that inquiry will refine the specific problem.

This process allows for innovating swiftly, broadly, and anywhere, even with limited resources.

Innovations solve problems

What you learn through innovation prototyping is to become an expert at the problem you aim to solve. You acquire that expertise by prototyping every facet of the problem—yes, prototyping the *problem*, not the solution—all the way down to the organization that will make the problem go away. Innovations in this process are described in terms of the problems. An innovation is the result.

Trial and Error

Trial and error is not "random exploration." It is because of this misunderstanding that the phrase "trial and error" has come into and gone out of fashion a number of times, and why some people want to turn it into an adjective that modifies "experimentation." In some contexts, phrases from the sciences—"hypothesis testing," "experiment," or "validation"—are used instead of "trial and error." They all mean very different things.

Trial and error reflects an extreme appetite for advancing through evidence. It begins with a suspicion that something might be possible, followed by an attempt to produce an "apparatus" to try out that suspicion. The objective is to make something work. If it does, you are done, though you might want to spend some time understanding what made it work so you can make it better. If it doesn't work, that is, there is an error, your job is to trace back what might have gone wrong. In the course of that discovery process you will formulate several hypotheses about why things failed. Together, these hypotheses will inform the next trial. You learn how to make your "apparatus" work reliably and consistently. The apparatus can be an artifact or an organization.

The kind of evidence one seeks through trial and error is broader than what can be encapsulated in a single hypothesis. The objective is to make numerous pieces work together in an intended way or to discover a new intention, unlike what you seek through hypothesis testing. The purpose of verifying a hypothesis is to acquire one bit of knowledge about an underlying theory. The tools we have for verifying hypotheses require that you develop and refute multiple counterfactuals before a single hypothesis can be verified.

The differences lie in the objective, in the amount of information, and in the subject of your inquiry. Hypothesis testing is an instrument to prove generality one bit at a time; the subject of your inquiry is a theory—knowledge. Trial and error is a means to make something work reliably by building an entire apparatus and letting your intuition guide the discovery process; the subject of your inquiry is what you want to build—something tangible.

You need only to make your innovation work. You do not need to prove the generality of the principles underlying your innovation.

Chapter 2 is about the problem at the center of your innovating. You can work with real-world problems at table scale, which allows for quick failure and broader trial and error. You can query and refine a real-world problem quickly and iteratively at that scale, using a set of parts to probe nature and inquiry to probe society. Your innovation prototype accumulates all prior knowledge about the problem to be solved, the "artifact" (whatever is created to solve the problem), the impact, and the vehicle necessary to bring the innovation to life. The innovation prototype is also a model of the uncertainties that remain. Along the way, you can acquire new skills, new team members, and new knowledge. You can learn how to scale up a solution, verify a solution, give a tangible demonstration of the problem, and develop an organization to meet the actual scale of the real-world problem.

Every example in this chapter shows innovating at work. It is how I've trained people at the Massachusetts Institute of Technology, across all disciplines, to innovate from the first moment they walk into class. Each of the innovators discussed in this chapter's examples had a hunch about a problem. That was their intuition, and it seeded their work. They operated with the problem at the appropriate scale to enable trial and error. They made their problems tangible through interaction with mostly existing parts and with a community. They learned to be productively wrong. Ultimately, each of their innovation prototypes evolved into a vehicle to scale up the solution to its community. Not one of the innovators was a subject expert on the eventual innovation. They were considered experts only after the fact. That potentially describes you.

Takeaways

The examples in this chapter illustrate five important observations about the path ahead for your hunch for an innovation:

- *Your hunch about a problem can lead to many innovations.*
At the beginning, there is an abundance of paths forward and potential outcomes. This book has no answer at the end. The "best" path is undefined (e.g., Ford, laser, Kinect, PCR).

- *No one thing is responsible for the whole.*
It is unclear which technology, part, or aspect of the organization was pivotal to an innovation's success (e.g., Greenpeace, PCR).

- *The path to an innovation is full of choices, not formulas.*
The technology, the market, the organizational vehicle by which the innovation reaches its community, and the form and magnitude of your impact are consequences of choices you need to make (e.g., PCR).

- *Innovating, like learning, is a highly nonlinear process.*
When the time comes to explain the account of your success, you will need to streamline your story and make it look linear. The full story of every trial, error, and corner turned is boring.

- *At its beginning, no thing about the eventual innovation is new.*

I've chosen stories that span social, technology, management, product, academic, and research innovation and range from startups to established corporations and social organizations. These stories convey that these observations are universal; together, the stories suggest a set of principles for innovating:

• A hunch about a real-world problem seeds the process.

• Working on the problem at scale enables experimentation through trial and error.

• You learn about the problem through interaction with mostly existing parts and with a community.

• Ultimately, the innovation prototype evolves into a vehicle to scale up the solution to its community.

• You do not need to be an expert at the problem yet. You become one as you learn.

There is a three-part checklist to get started innovating:

1. A hunch about a real-world problem.

2. A set of parts and access to a community of people to render the problem tangible.

3. A strategy to engage in trial and error, and an appetite to learn by being wrong first.

You *learn* about the problem as you bring together people and parts.

INNOVAT**ING**

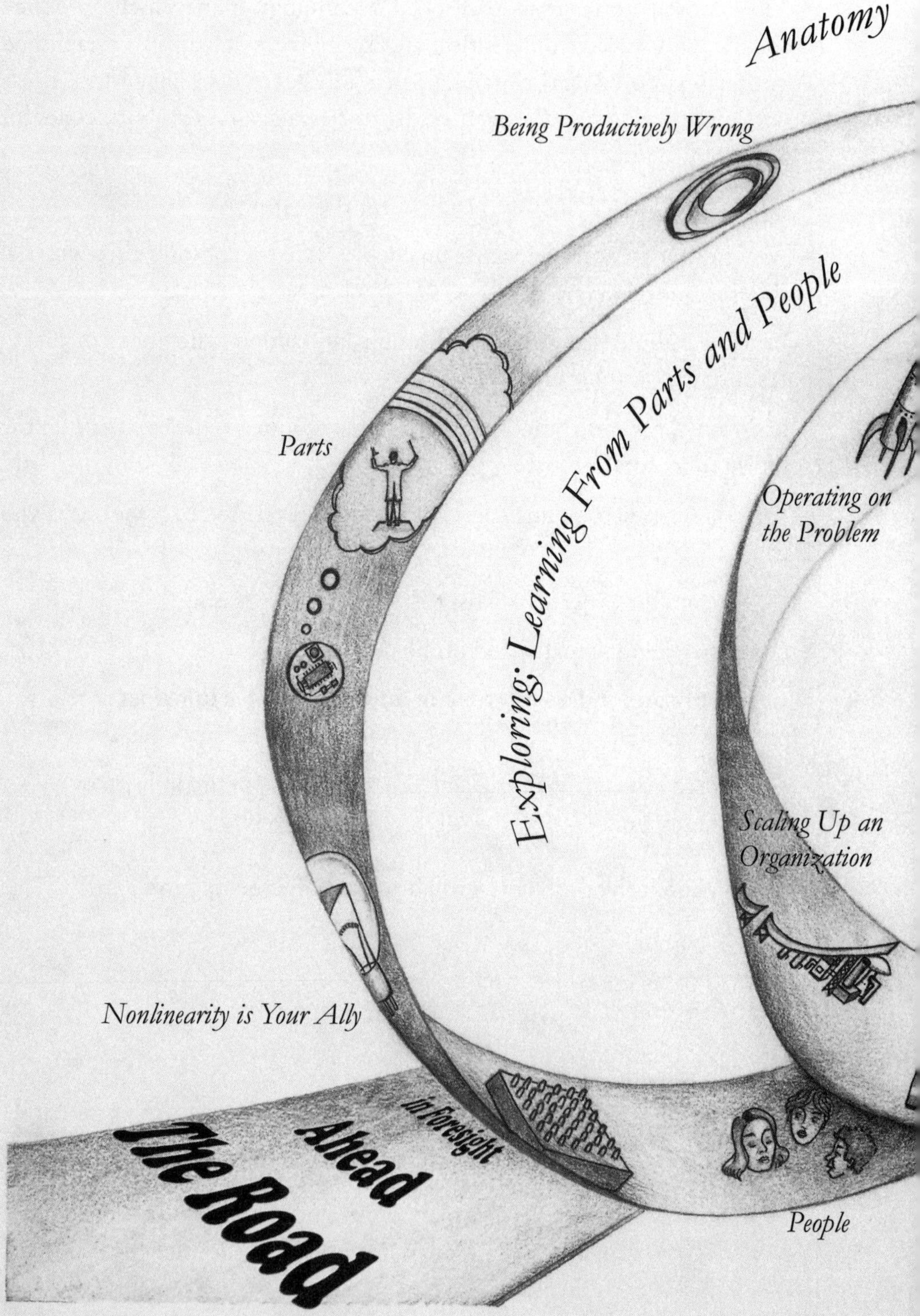

Anatomy

Being Productively Wrong

Exploring: Learning From Parts and People

Parts

Operating on the Problem

Scaling Up an Organization

Nonlinearity is Your Ally

People

in Foresight

Ahead

The Road

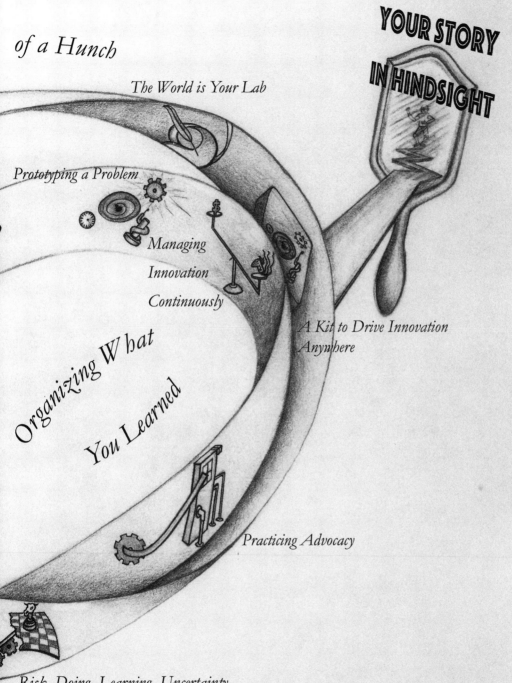

of a Hunch

YOUR STORY IN HINDSIGHT

The World is Your Lab

Prototyping a Problem

Managing

Innovation

Continuously

A Kit to Drive Innovation

Anywhere

Organizing What

You Learned

Practicing Advocacy

Risk, Doing, Learning, Uncertainty

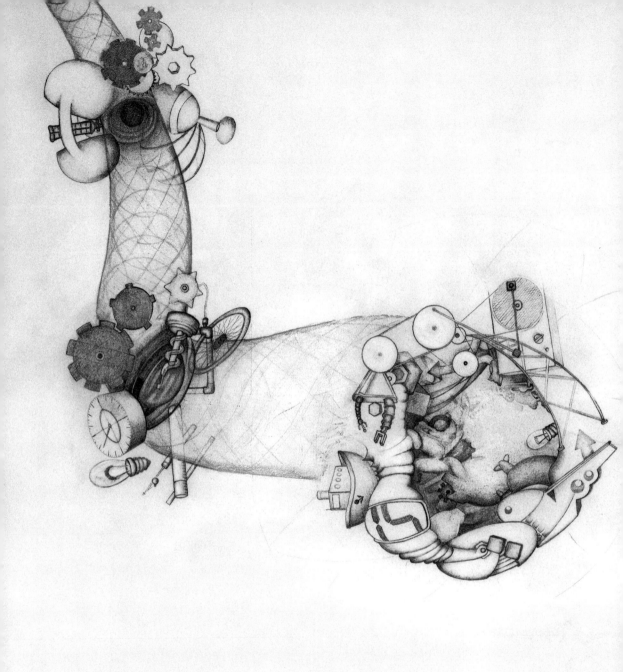

PROTOTYPING A
REAL-WORLD PROBLEM

Everything you consider an innovation came to your attention because it had an impact. Specifically, it solved a real-world problem.

At first, postulating a real-world problem may seem as formidable a task as coming up with an earth-shattering idea. However, the problem does not have to be big; it just has to be real. Indeed, the examples in chapter 1 suggest that all you need is a hunch about a problem. You'll refine the problem as you innovate. You'll innovate as you solve the problem. The subtlety lies in appreciating that there is a pattern to how we think of real-world problems that have been solved and making it routine—practicing—to give your hunch the same structure.

By the end of this chapter you should be able to persuade yourself that a hunch about a problem—not yet a full-blown problem—is the most uncomplicated starting point for an innovation. You can start there repeatedly without fear of being wrong. In fact, there is a fairly unassuming way to build a purpose from a hunch that will lead you to an increasingly well-defined problem.

A hunch about a problem is a realization that there's something that's not quite right and that you suspect you can change. The founders of Greenpeace seem to have been driven by their perception that "the planet was in danger"[1] and that society needed to be made aware of that through action. Your hunch may emerge from something in your daily life, from something in your work environment, from recurring conversations within your community, from something you wish were true, or from a particular problem you experience that you suspect could go away if something else existed—possibly something you can't quite formulate yet.

Hunches

There is nothing particularly enlightening about a hunch. In most cases, a hunch is just a figment of an idea that comes to you from conversations in your community or from something you see in your surroundings.

Students routinely come to class with a "great idea" to do something "new." Almost without exception, they claim that no one else has ever thought about their idea. That's the first sign that they are confusing novelty with unfamiliarity.

Later in the semester, they discover they were wrong; someone else *had* thought about the same thing. And because they had banked everything on the presumed novelty of their idea, they feel compelled to abandon it.

That's not how a real hunch works.

As the stories in chapter 1 show, no thing is new at the genesis of an innovation; rather, novelty accrues as you innovate. The people who lead their hunches to impact developed those hunches through involvement in a community, an interest group, their own life or work experiences, or from some event or news related to a personal interest that they chose to act upon. In a way, hunches seem to brew over time through the exploration of your own interests. You may want to take that as decidedly good news; it means you've probably already brewed a few hunches of your own. For now, no magic is required.

That said, here are some examples of hunches from the stories in chapter 1, as well as some that have emerged from my research and my experience as an educator:

"The planet is in danger."

"A billion people have no access to safe drinking water."

"Gas prices keep going up."

"We need to become energy independent."

"Electrical energy is lost in transmission and distribution."

"We want to make the Internet intelligent."

"What if we could turn Science, Technology, Engineering, and Math (STEM) into a game and an activity for parents and kids to do together at home?"

"I want to build the best computer possible for myself."

"My files are never with me when I need them."

"We just invented a new kind of battery, so let's do a startup for cellphone batteries."

"If we could draw water from the air, we wouldn't need desalination plants."

These examples simply tell you that coming up with a hunch is easy. Your inquiry may begin with a fact, or even a syllogism. It need not be correct. It need not even be physically viable yet. That's why you are allowed to be wrong.

A *hunch* about a problem is the most uncomplicated starting point for an innovation.

You ought to give that hunch the same structure problems have. In fact, all you need is deceivingly contained in the statement "every innovation solves a problem."

That statement may feel like a circular reference—a truism—that will become trivially true at the end of your process. If you work backward from hindsight with some of the examples from chapter 1, you may appreciate that the problems those innovations solved are seldom defined directly. Rather, they are defined indirectly: by a solution such as "the affordable car"; by how a solution empowers the members of a community, such as the description of PCR as a means to "amplify a genetic signal"; or by how an organization goes about verifying that it is indeed addressing the problem, such as "Greenpeace ... uses non-violent, creative confrontation to expose global environmental problems, and to force the solutions which are essential to a green and peaceful future."[2]

Innovations solve real-world problems.

Using these stories for guidance, here is how it might play out for your eventual innovation: An organization (new or existing) will serve the outcomes of your innovating sustainably. Let's call that your "offering." You'll describe those outcomes as a solution that negates a problem. You'll likely illustrate a form of the problem, indirectly, by the benefits that emerge from adopting your offering. With guidance from your organization, the receiving community will infer the specific instance of the problem your offering solves for them; the benefits of your offering will persuade others that the problem is indeed solved and that your solution creates value[3] for them. And you'll see a rising number of players building from what you have to offer, some extending it, some copying it.

Coming up with an explicit problem statement is difficult, and, as these examples suggest, you may never need to have an explicit problem statement like those common in engineering and mathematics. Rather, your problem may be best illustrated by the benefits of your innovation and the organization that makes it happen. Explicit problem statements are, as a matter of fact, an element of your story in hindsight. I have become convinced that it's impossible to define a problem explicitly in foresight. The best we can do is define a problem indirectly, and that's why I'm having you focus on a hunch.

Early on, however, before the opportunity for innovation is apparent, the need for specificity drives many of us to try to use various formalisms to describe a problem. A common one is "*N* people in place A cannot do X and need Y." Say, for instance, "783 million people do not have access to clean water."[4] That, in a way, is a prototype of the magnitude of the impact we hope a solution to some unstated problem will have, but it is not a problem description. It isn't even really a problem. It is, potentially, a statement of fact. It can be a hunch.

We can, however, define the attributes of solved real-world problems. Still looking at the stories in hindsight, the statement "every innovation solves a problem" and the historical examples I've overviewed imply three things about the kinds of problems innovations solve. They are:

- *solvable*: At least one solution to the problem exists, and probably more.

- *recognizable*: The problem, the solution, or what the solution must achieve can be made tangible.

- *verifiable*: There is a way to decide whether a solution indeed solves the problem.

In logic, the word *decidable* describes problems of this kind. In the context of innovation, you may think of it this way: Eventually, there will be an organization that can follow a set of steps to serve a solution to the problem repeatedly and sustainably; the members of the receiving community will be able to recognize rapidly that what the organization serves does indeed solve the problem. This, and specifically the third condition, can feel so obvious that you may overlook it altogether.

But it is important. Certifying or verifying a solution should take significantly less effort than it took to solve the problem; otherwise, future beneficiaries would have to spend as much time as you did to figure out how what you produced solves their problem.

A Medieval Cell Phone

To illustrate this issue of others determining whether you've solved a problem of theirs, let's do a thought experiment. Take your mobile phone back to the Middle Ages, along with another phone, a cell tower, and a generator, and show some decision makers—members of the aristocracy, clergy, and so on—what your phone does. Let's say you survive: Suppose you are not burned at the stake in the first five minutes, and they actually pay attention to what you've shown them. Can you even fathom that they will be able to agree that your mobile phone solves a communication problem?

As you prepare your argument, you might want to consider what Sir William Preece, chief engineer of the British Post Office, said about telephones when they first appeared in America, around 1878, quite a number of years after your most certainly failed attempt to convince anyone living in the Middle Ages that the phone solves a problem:

> There are conditions in America which necessitate the use of such instruments more than here. Here we have a super-abundance of messengers, errand boys and things of that kind. … The absence of servants has compelled America to adopt communications systems for domestic purposes.[*]

[*]Marion May Dilts, *The Telephone in a Changing World*, Longmans, Green and Co., 1941. For additional quotations from people attesting to the uselessness of the telephone, see "Imagining the Internet: A History and Forecast" at http://www.elon.edu/e-web/predictions/150/1870.xhtml.

Giving your hunch the structure of a problem

You may take these attributes of solved real-world problems as principles to start working on your hunch. The problem that gives you purpose—regardless of how elusive it may still be—must satisfy three simple conditions. You ought to be able to:

- imagine one or more specific solutions,
- describe the problem or what a solution to the problem must achieve, and
- verify when the problem is solved.

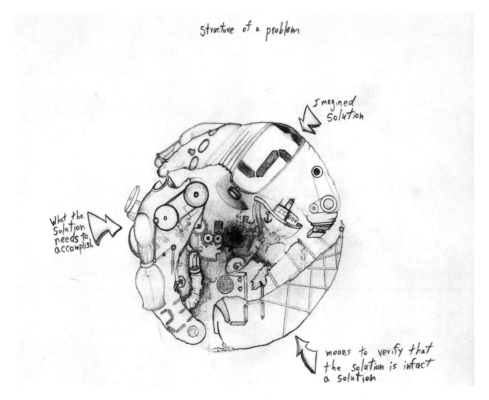

Three conditions that must be met to define a problem. The concepts illustrated are partially inspired by the examples given in chapter 1. The "black hole" at the center signifies that problems are seldom defined directly.

The Theory of Computation

If you are familiar with the theory of computation, you may recognize the notions of *recognizable* and *decidable* I use to give structure to a problem. I am drawing an analogy between a smoothly working organization and a Turing Machine—a concept invented by Alan Turing in the 1930s that can be adapted to simulate the logic of any computer algorithm and is particularly useful to reason about the execution of processes.

An idealized organization will take some input (skills, raw materials, knowledge, parts, support query, user feedback, etc.) and its business units will perform numerous actions, following some sequence determined by the input and by the organization's structure. The sequence will continue again and again until the idealized organization delivers the result of its operations to a final user. That's the end state.

The analogy goes only so far, however, because unlike Turing Machines, organizations—even process-based ones—have to deal with uncertainties and with a constantly changing environment. But it is useful to think about the problem the organization you have to build must solve, and to think about the task at hand.

In a way, your task is to rephrase the problem you were given so it has the characteristics of the types of problems that can be sustainably solved by an organization following a finite number of steps, and to devise the organization that can successfully and reliably produce a solution. Were this actual computer science, your task would be to understand the structure of the problem so you could build an algorithm to address it.

In your quest, you may need to invent and innovate on many fronts so this becomes possible. That's why you can start with things that are not new and still come up with an innovation at the end.

You may be wondering why you need an organization at all. The problem you are solving, whatever it is, will disappear only if the solution is served to the community experiencing the problem. Simply *sharing* the solution with people rarely works; more often, you need to develop and supply the solution—and that is why you need an organization. It may be two people in a basement, a research lab, an educational endeavor, or a company that employs thousands around the world. The kind of organization you develop and how long it needs to run is a choice.

Your first hunch will most likely address these three conditions only partially. As a matter of fact, hunches, at first, are either incomplete or wrong about several aspects of the problem. Most often the imagined solution, "the product," is wrong—so much so that making it a purpose to identify the ways in which you are wrong may prove to be the most successful strategy. Being wrong can be detected easily because at first, your best guesses for the three conditions will likely contradict one another. That's what prototyping and, more generally, trial and error are for: to resolve those contradictions.

These three conditions build a purpose around your initial hunch that has the structure of a problem like the ones successful innovations solve. There are advantages to focusing on something that has the structure of a problem over focusing on a technology, a product, or a startup idea—even when the specifics of the problem are still elusive, when the technological idea is pure and you still believe it is capable of solving every problem, or when you are convinced your imaginary product is perfect.

Working with these three conditions helps keep you honest. For instance, it is more difficult to fall in love with that product you imagine if you must at the same time persuade yourself that any solution to the problem you imagine must achieve a specific set of things dictated by the problem, not the product. Ditto for a technology.

As you innovate, your purpose is to make the problem increasingly clear and specific. Focusing on the problem will also help you find a way to apply the same kind of mental operations we know work for problem solving, which you began to learn intuitively as a child and have been practicing ever since. Finally, these three conditions give you something to do right away. No matter how tenuous your hunch, each of the conditions above gives you a way to strengthen it by making the problem tangible.

Structure of a Problem

In workshops, I have used hunches to develop broad problem statements to get students started with innovating. Here's an example of a hunch for energy independence, structured as a problem, I have used with workshop participants:

> *How do YOU become energy independent? How will you know? Can you generate and distribute power reliably and locally for your home? How about your neighborhood?*
>
> There is a lot of talk about the smart grid, variable sources of generation, and the role that renewables and alternative energies play in homeland security. Solving the problem of an entire country is "easy" with some policy and economic alchemy, although that's beyond the scope of this workshop.* But what about your energy independence? Simple inspection of your electrical bill can show you that a significant portion of your domestic security (as measured in $) goes to ensure transmission and distribution of the energy you consume. Solving this problem at the scale of your home or neighborhood requires addressing questions about smart combination of variable energy sources, predictive technologies and storage. While this may make the problem tractable/analyzable at a lab scale you may also wonder:
>
> Can this be scaled up?
>
> Can you regain a country's energy independence one neighborhood at a time?

Here is how the initial statement maps the three conditions.

• It guides you to *imagine specific solutions* in the electrical space. You could imagine anything from installing generation sources at your home to serving an entire community or from managing distribution with smart technologies to introducing storage. These solutions, though, begin at a local scale.

- It *describes the problem in terms of what a solution must achieve*, by linking energy independence to your security, as measured in dollars—not unlike how governments link dependence on external energy sources with national security. However, the application is constrained to you. That is, you describe the problem and solution at an individual or community level.

- It suggests several *ways to verify is problem is solved*—for instance, a measure of the recurring dollar amount spent in energy as compared to the rest of the budget and to the control you have over that dollar amount.

This example shows you how easy it can be to give a hunch the structure of a problem. Further, it stresses that these steps are not hard to execute. In fact, they are designed to give you just enough structure to aid your creative self. What makes innovating interesting isn't how formidable an inquiry you set for yourself or how complex a problem you set out to solve, but rather what you bring to the picture—that is, your critical thinking about the specifics.

People often believe they must somehow dress up the problem they are targeting to make it "inspiring." However, the more down to earth a problem, the easier it is to start applying your intelligence to solving it. Problems have an inherent complexity, and solutions—once real—are *always* inspiring. Simply put, it is a waste of time to make things look or actually be more complicated than they really are. I find it much more useful and meaningful to apply my brain to unraveling the real complexity of the problem, not some artificial complexity I invent to impress others or garner attention.

*Thanassis Cambanis, "American energy independence: the great shake-up," Boston Globe, May 26, 2013. Accessed May 9, 2015. http://www.bostonglobe.com/ideas /2013/05/25/american-energy-independence-great-shake/pO9Lsad4cVQvjdpyxMI1DO /story.html.

Making the problem tangible

You begin to understand the problem by working on it—either the solution, what the solution needs to achieve, or the means to verify your solution. Let me show you by example each of these three ways to work on your hunch.

Working on the first condition—*the solution you imagine*—is the most readily apparent option to make your problem tangible. Each of the examples in chapter 1 involved an early prototype of an imagined solution. The prototypes history has recorded may not have been the very first ones—many of them worked, and hardly anything works as expected the first time you try. But they are early enough to suggest a given about first prototypes. Each used readily available parts and demonstrated or enabled aspects of the eventual innovation. Henry Ford's early "quadricycle," for example, demonstrated aspects of the affordable car and enabled Ford to move forward; he eventually developed the assembly line and the financing innovations that made the Model T affordable. Dr. Maiman likely realized how to pump light using a flash lamp over an earlier incarnation of his laser design. Johnny Chung Lee's website of projects showed how to produce several fully working systems. In my telling of the Greenpeace story, I call the Amchitka trip an innovation prototype, but billboards and the boat that sank are both early prototypes.

If you decide to make the problem tangible by working on the solution you imagine, you do not need to spend time worrying about whether your solution is an innovation. With the examples I shared in chapter 1 as guidance, you have to be ready to accept that your innovations will emerge along the way.

You can also work on your hunch by *simulating the problem or what a solution must achieve*. This may appear to be a distraction when you are focused solely on a product. It isn't a distraction when the objective is to understand the problem.

Problems that seem difficult to grasp may be easy to "simulate." For instance, in August 2013, several newspapers reported on Brendan O'Connor's demonstration of the problem of wireless security and privacy. O'Connor—a security researcher and a law student—reportedly wondered how easy it would be for a private citizen to monitor the movements of everyone on a street. At a cost of $57 per box, he put together a bunch of parts—some plastic cases, some credit-card-size computers, some inexpensive sensors and Wi-Fi adaptors—that he connected to a command-and-control system, and then created a way to visualize all the wireless traffic his system would pick up, including smartphone traffic.[5] He named the box CreepyDOL and built ten such boxes. He presented his system at several security conferences.

O'Connor's purpose, as reported by different news agencies, was not to deploy the system for espionage. The system he built can be seen as an illustration of the obsolescence of privacy protection laws at the time. It shows the ease with which private citizens can spy on each other's digital lives, and it probably also helped him market his services as a security consultant. It is a prototype of the problem, not a solution. By making the digital privacy problem tangible, CreepyDOL specified implicitly many of the things a solution would need to achieve without ever needing to propose one. It also provided a tangible platform for further exploration.

> You innovate by making the problem increasingly clear and specific.

Finally, you can work on your hunch by assessing strategies to *determine the problem is solved*. This may seem counterintuitive at first. In truth we do it all the time, but think about it only after the fact: For instance, most established businesses have quality-assurance practices; government bodies have certification and compliance tests and criteria. Even mathematical problems are phrased as "find X," where X is generally specified via a set of conditions that the unknown has to satisfy; problem solvers are encouraged to find ways to verify their solution. These are all strategies to verify that the candidate product or solution meets what's expected. When both problem and solution seem too elusive to get you started, you may still be able to begin by working on a "verifier." The idea is simple: If by happenstance you were to stumble upon a candidate solution, how would you verify that it indeed solves the problem?

The X-Prize Foundation model to develop prizes offers several examples of "verifiers." The first X-Prize, later named the Ansari X-Prize, was conceived around 1995 with the vision to "make space travel safe, affordable and accessible to everyone through the creation of a personal spaceflight industry."[6] That was the problem the foundation wanted solved—or, more specifically, traveling to space was the problem.

Rather than specifying a solution or working on a solution, the X-Prize Foundation worked on specifying what a solution would need to *achieve*. The aim of the X-Prize became persuading people that a private space industry was possible and empowering entrepreneurs to get that industry started.

> Abandon the expectation of a linear process so you can progress by learning about the problem you've set out to solve.

The winning team would "build and launch a spacecraft capable of carrying three people to 100 kilometers above the earth's surface, twice within two weeks." In 2004, Spaceship One did just that and won the $10 million award.

The statement "build and launch" is, in fact, a verifier—a tool to determine the challenge had been met.

These examples illustrate general strategies to make virtually any problem tangible. The examples show tangible demonstrations of problems that range from products, to new industries, academia, policy, and law, to name a few. They show what I mean by prototyping innovations: rendering your problem tangible so you can bring together the knowledge and skills you have and the knowledge and skills you will acquire toward the development of an organization that will ultimately solve the problem. You learn and innovate as you go.

Innovating and problem solving

There is a lot of work to do from the initial hunch until an organization that sustainably solves a problem is in place. Along the way, it is very easy to become enamored with intermediate steps or pose difficult questions such as "When should I stop ideating?" or "When should I start executing?" In other words, it is not uncommon to ask questions that implicitly assume there is an "answer at the end of the book." Not only is that not the case, but it's also an unproductive way of thinking about innovating. It is a trap. While there may seem to be some comfort in seeing the process of innovating as having specific, definable stages governed by "progress indicators," you benefit most when you abandon the expectation of linearity and liken progress to learning about the problem.

Giving your hunch the structure of a problem protects you from falling into the trap. It shifts your mind-set away from certain isolated aspects of your innovating, such as the organization, or the technology, or the product, or the design, or manufacturing, or its placement; it makes them all auxiliary elements to the problem. Nothing is "pretty" until the problem is solved, nor does it need to be.

The characterization of innovations in terms of the problems they solve makes innovating akin to a very general kind of problem solving in which both the problem and the solution are discovered as part of the process.

The problem that will give you purpose is unlike mathematical, scientific, engineering, or even market problems. Such problems generally come specified by a well-defined set of parts, or by a reasonably well-defined endpoint. The kinds of problems we seek to address with innovations, whether inspired by the market, society, or technology, are fairly sophisticated constructs that may at first lack any definition—everything, even the endpoint, is a choice you make.

In his book *How to Solve It*, first published in 1945, George Pólya describes problem solving formally with a list of mental operations.[7] He introduces them as questions teachers and problem solvers may pose to themselves to aid progress in resolving a problem, as well as to uncover signs of progress when progress is not readily apparent. Pólya groups the questions into four phases: understand the problem, make a plan, carry out the plan, and look back on your work to see whether you could solve the problem in a better way.

Pólya was writing about mathematical problems and was extrapolating to practical problems that have a clear endpoint. But problem solving can be generalized to the broader problems you encounter when innovating. This structure—solvable, recognizable, and decidable—establishes for an innovation problem the same structure as mathematical problems. So, in principle, you should be able to begin making your problem clearer by posing to yourself the questions Pólya offers—something you'll have to do at several points along the way. However, because the data, the unknowns, and the conditions that define your problem are also part of what you need to solve for, you are going to need more to go on.

Pólya provides numerous recommendations of ways to overcome difficulty or lack of progress when solving a problem. For example:

- "If you can't solve a problem, then there is an easier problem you can solve: find it.

- "If you cannot solve the proposed problem, try to solve first some related problem. Could you imagine a more accessible related problem?"

- "Draw a hypothetical figure which supposes the condition of the problem satisfied in all its parts."

These three recommendations provide a good basis for generalizing problem solving to innovating. The first thing to realize is that the problem you are proposing very likely lives at a scale far beyond your immediate resources. So, you need to find an easier or more accessible version of your innovation problem to solve. Put another way, you need to scale the problem you want to solve down to a scale at which it corresponds to the resources you have to understand the problem.

For a mathematical problem, a figure would help you turn something otherwise abstract into something tangible that helps your intellect and senses work together. Blueprints or models serve the same purpose for practical and engineering problems. For entrepreneurship and innovation, you might use a slide-deck. But that can help you only so much; a lot remains abstract. If you've brought your problem to a resource-friendly scale, though, there are other things you can do to realize the same value that figures offer in solving mathematical problems. You can physically prototype any aspect of your problem, or, as Pólya puts it, you can build a prototype that "assumes the condition of the problem satisfied in all of its parts." That, of course, might include a gizmo, but it can also include an organization, distribution, marketing, manufacturing, and so on. At that scale, you don't have to limit yourself to prototyping form alone. You should strive to prototype function.

So, you can generalize problem solving to innovating if you work on bringing the problem first to a resource-friendly scale, and work at that scale to make the problem tangible, prototyping all aspects of your eventual solution—and, in so doing, implicitly outlining all areas in which innovations may be required. After that, your task will be to understand your problem at one scale and work toward scaling up successive demonstrations of the problem.

Scale

The objective of bringing a problem to a different scale is to enable quick and tangible experimentation on the aspects of the problem most critical to move forward. It is also helpful to begin separating the nature of the problem from the magnitude of the impact to which one aspires.

There are many ways to work on the scale of a problem. A common one is to change the size of the community—for instance, "Let's start with five people and then ramp up to twenty." But there are other, more effective ways to bring a problem to a scale that is more amenable for quick experimentation—for instance, introducing assumptions to extract the most knowledge and impact from the resources at hand.

Let me give you two examples of the interplay between resources and scale.

In class, a group was interested in devising a system to detect infectious diseases quickly. To realize their initial vision, they would have needed a lab with biosafety level 2 or 3. The stage their hunch was at, though, did not justify the investment in resources and skills required to access and use such a facility.

They could have stopped at that. Instead, they introduced an assumption of scale: a strawberry is a bacteria. The effect of this particular assumption of scale was (a) because of how easy it is to extract DNA from a strawberry, they could forgo dealing with the complexity of establishing what constitutes a good enough sample of bacterial DNA for their experiments; (b) they didn't need the extra resources of a biosafety lab right away; and (c) getting out of the straitjacket of thinking in terms of the biosafety lab freed them to think about a simple device with which to experiment on the problem. After that, it took only a week to bring together the parts and knowledge they needed and come up with a plan for how to move forward. They were able to articulate their plan by building on conversations with industry experts and a demonstration of a small working device.

More generally, working on the scale of the problem also introduced a very convenient change to the sequence of proofs of concept. Once their device was ready, the knowledge they would need next would transcend their assumption of scale. At that point, they would need access to the specialized lab only to test for that specific knowledge, and they could either contract out the testing or rent the lab whenever they were ready to move forward.

In a different setting, during a lecture in which I challenged students to think about how to prototype their ideas (form and function) all in one day, a team complained that their idea could not be prototyped. They were thinking about a pill that would emit a signal when dissolved in the stomach and help measure patient compliance. Their main concerns were miniaturizing the electronics to fit in a pill, any regulatory unknowns that might exist, and what a safe signal strength would be.

On that occasion, the answer to the question of scale was to assume a much bigger human. In other words, miniaturizing electronics and inserting them in a pill were considerations for down the road—considerations that might indeed require significant innovations. At that day's stage of their thinking, though, the team needed to *characterize* the problem. Coating the necessary electronics in cereal, choosing a recipient of the right size to simulate a stomach proportional to the size of the pill, coating the simulated stomach with material that had a density similar to that of a human body, and then developing a number of test scenarios would at least give their questions the next level of specificity they would need to outline all the manufacturing steps, regulatory measurements, and reasoning about the mechanisms by which they could actually measure compliance.

You can generalize problem solving to innovating: Bring the problem to a resource-friendly scale, make it tangible, and scale up successive demonstrations of the problem.

The final problem you'll solve will likely differ substantially from the problem you thought you were solving at first. That is because your first expression of the problem was, with high probability, ill informed if not outright wrong. That's all right; the purpose of your first hunch was to get you started. You'll discover how wrong as your innovation prototype evolves toward scale.

Prototyping the problem at scale

With your hunch structured as a *problem*, your next task is to make that problem tangible at a scale that matches your available resources. You can do so by prototyping a solution, the problem itself, or even the means to verify the solution—just as was done in the examples of Ford, Greenpeace, CreepyDOL, and the X-Prize. Only as you make progress in your understanding of your problem does it become worthwhile to address it at a larger scale. That's when you'll know whether you might need additional parts, people, and resources.

Progressing from a Hunch

1. At the start, you a have hunch.

2. You give your hunch the structure of a problem. You have to postulate what makes your hunch solvable, recognizable, and verifiable.

3. You reformulate your problem at a smaller scale so you can start working on it.

4. You sketch the smaller-scale problem with a prototype of the problem's form and function, much as you would draw a figure to aid in problem solving.

5. You can now describe your prototype as a small-scale illustration of a problem.

6. You restate the problem.

Let me put this more pragmatically as a set of actions that might help you progress from a hunch to a full-scale realization of a solution to a real-world problem inspired by your hunch.

At the start, you a have hunch. You give your hunch the structure of a problem. You need to postulate what makes your hunch solvable, recognizable, and verifiable. The key idea is that you need to configure your endeavor in a manner that allows you to probe the specifics of the problem you think you are solving. Doing so makes you aware of all the things you do not know just yet about the problem, which generally come in two flavors: things you can address now and things you will have to address later.

There are things you don't know because you have never studied or learned them, but you can at least recognize. These are data or conditions that will further define your problem. You need to find out about them. Someone else might know.

Then there are things you can't recognize or can't yet formulate. Perhaps no one has ever formulated them. These are your unknowns and uncertainties. You need to find the scale for your problem at which they are no longer an issue or can be assumed resolved. By the time you hit the scale at which they do matter, you'll be able to reason about them as data or conditions someone else might know about.

Problem structure is also addressed in chapter 6, where I discuss how to set up a problem you care about so a team can work on defining, refining, and ultimately solving it.

You reformulate your problem at a smaller scale so you can begin working on it. This action is a response to a very straightforward and all-too-common question innovators face: What do you do when the problem you want to solve is at a scale far beyond your immediate resources?

You have options. One is to "prototype" your hunch with PowerPoint, investing resources you *do* have, including your time, to create a presentation as part of the quest for further resources.

An alternative is to reformulate the problem so you can begin working on it. Doing so establishes, for now, a good endpoint from which to work backward with the resources at hand. Scaling down teaches you what you will need to deal with—namely, those unknowns and uncertainties—as you scale up in the future and layer proof of concept upon proof of concept.

You can prototype a problem with just about anything that can show you how you are wrong—such as parts and people, covered in chapters 3 and 4.

You sketch the smaller-scale problem with a prototype of the problem's form and function, much as you would draw a figure to aid in problem solving. Your prototype of the problem, however, will have to assume that the problem is solved; that is, it will have to contain all the parts you expect a solution to have. Your goal is to prototype both the parts and the impact. Successful prototypes inform, in one way or another, everything involved in making the future offering real. An innovation prototype is a bit like a dry run of your entire idea.

If at this point, if your idea of an organization is still fuzzy and you are uncertain about what to prototype, you need to read the problem statement you've elaborated from your initial hunch. You can prototype the solution, although prototyping a solution when the problem is still elusive can be stressful. It might be easier to begin by prototyping something that answers questions such as "What needs to be true?" or "What am I trying to accomplish?" or "How will I know when to stop?"

You can prototype a simulation of the problem. You can prototype something that will allow you to verify that a candidate solution is indeed a solution. All there is to work with are parts and people. Your task at this point is to figure out which parts and which people to bring together.

The parts you bring together help you materialize data and facts you think you know about the problem. Those parts may include a gizmo as well as aspects of the organization you think can serve a solution sustainably. People help you address the things you do not know (data and conditions).

At first, your main concern is what to bring together. The data, unknowns, and conditions take over your inquiry. The task changes as you begin to worry about how to make the parts and people work together, and ultimately it becomes a matter of how to do so effectively, which is defined as you progress in the problem. I discuss this further in chapter 7.

You can now describe your prototype as a small-scale illustration of a problem. By now, you have made certain assumptions to make the initial problem tractable and have discovered what else needs to be true to bring it to the next scale. The version of the prototype you just completed includes the following:

- the data you gathered

- parts that have been "strung together"

- a group of people (either colleagues or people you have accessed for information, feedback, or knowledge)

- a rough vision of impact, possibly even a small-scale demonstration, and a sense of what else needs to be accomplished and the uncertainties that remain.

Your prototype is all those things together: the artifact you put together, the people you probed, the people who joined you at this stage, the people that already left the group, the knowledge you've acquired, the impact you've prototyped, the larger problem you envision, and the emerging organization that sustains your inquiry into the problem. Some things in the prototype worked; some did not. Some you omitted because you either don't know how to address them yet or you lack the resources to address them now. Perhaps you think they can only be addressed later.

You can describe the problem by pointing to the aspects of the prototype you will work on next to make the imagined impact a reality. Describing the prototype of a small-scale illustration of the problem allows you to outline all that's known and the uncertainties that remain, and to rank order the resources you need by how they will help you progress. And with that, you could now advocate to obtain those resources, whatever they might be: funding, colleagues, time, expertise, space, and so on. I discuss uncertainty further in chapter 9.

You restate the problem. Back to square one? Not quite. Problems evolve. That's the message of the Greenpeace example. In each iteration, the problem gains resolution and you progress toward a larger-scale appreciation of the problem. Innovating is not the mystery it is sometimes made out to be; in fact, anyone can do it.

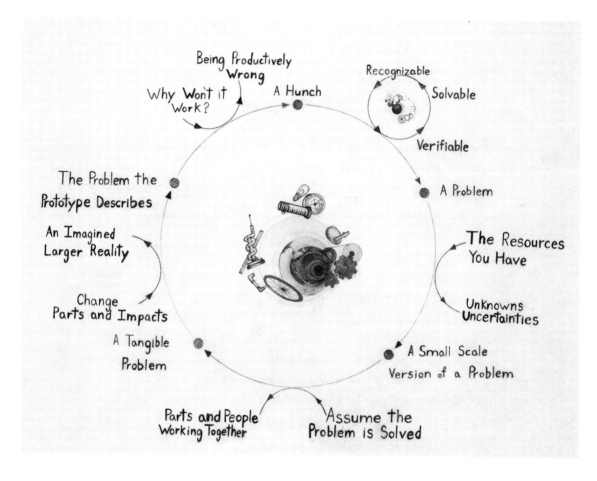

Visualization of the recurrence rule to refine a problem.

Iteration

Strictly speaking, the set of actions that I propose for progressing from a hunch may be interpreted as an iteration. The outcome of one pass through all the steps brings you back to the first step, which is a new beginning, but with a different understanding of the context in which you are working. Nevertheless, I avoid using the terms "iteration" or "iterate" because they tend to be confused in normal use.

Outside of computer science, "iterate" is often used to mean repeating something with the purpose of refining some answer. Other words from computer science, such as "variation," "relaxation," or "approximating," would be better choices to describe that process. They all, however, imply a sense of gradual convergence, which when absent makes it difficult to recognize progress or imagine the scenario that would lead one to change course.

When used in the non-technical sense, "iterating" often refers to a sequence of steps that end at a point different from the beginning. Also, within the process of iteration as the word is typically used, there is not usually an admonition to articulate the state of all the variables at the end (of each iteration). So, iterators are left with endpoints that resemble nothing like what they started with. Sometimes "iterate" is used even more ambiguously to mean that you restart or that you will refine whatever you are working on in some unspecified way.

I see the typical approach to iteration as a kind of insanity, because repeating the exact same steps should lead to the same outcome. To resolve the conundrum, people invoke learning without defining what that is—which makes it even harder to apply the process. So I try to avoid using the word 'iterate' altogether.

With your prototype in hand, with everything you have learned, and with everything you assumed, you should now realize why your first version of the problem was wrong. That's your next hunch. You may now restate the large-scale problem. You'll probably realize there isn't one but several versions of the problem you could focus on. Pick one for now. What is a possible solution? What does the solution need to achieve? How would you know the problem is solved? That's your next problem.

To be productively wrong, you have to make sure you can continue to execute. At some point, that may imply searching for additional resources. It is generally easier to search for those resources when a problem is tangible. Chapter 8 provides an overview of strategies for inquiry and advocacy with an innovation prototype.

The set of actions described in this "Prototyping the problem at scale" section will keep you busy as you bring your hunch to full scale. It is not a recipe, though. It is a progression of activities that ought to help you further refine a problem for innovation. You'll likely go through this progression multiple times as you learn about the problem you are solving. Often, as you try to restate your problem, you'll find yourself not with one but with multiple possible problems to pursue. That is actually an indicator of progress.

At first, the problem you solve will evolve substantially. That's because while the real-world problem that inspired your hunch remains largely the same, you are still discovering the actual content and characteristics of that problem. To an outsider, it will seem as though you are continually changing direction. That's mostly because what an outsider has to measure is your current best guess of the nature of the impact, your product, the organization, the technology, and so on. Those, though, are what Pólya might call auxiliary elements of your inquiry about the problem.

As you gain clarity on the problem you are solving, the same activities will gradually lead to a different outcome. It will increasingly look as though you layer proof of concept upon proof of concept, each demonstrating a larger-scale version of the problem and your proposed solution.

This should not come as a surprise: Numerous disciplines and industries have standardized the processes by which solutions demonstrate their readiness for scale. The progression from pre-clinical through various phases of clinical studies before approval of a new compound as a drug is one example. For new chemical and biochemical processes, the progression from lab scale to pilot plant and to commercial plant is another example. The same logic may be found in several other contexts, such as the development of manufacturing processes, and even with software as it progresses from beta to alpha. The specific meaning of scale varies in each context, but the underlying rationale is the same: Once the solution is clear, assembling all the components required to serve the intended population sustainably and reliably requires progressing through a scale-up sequence in which problem solving is key and there is no closed formula demonstration. The objective in these instances is, generally, called "de-risking."

As you gain clarity on the problem you are solving, the same activities will gradually lead to a different outcome.

Evolution of the Problem Greenpeace Solved

The problem that gives purpose to an emerging organization evolves as much as what the organization does. This can be seen by reviewing statements Greenpeace has used over the years to describe its purpose.

The progression of Greenpeace's statements illustrates that the problem innovators solve evolves as they innovate. From the outside, it appears that organizations and people pivot. This evolution, however, may also be explained by the difficulty of matching the richness of detail in the story the founders of Greenpeace lived going forward with the abundance of information about the organization and its context available when we retell the story in hindsight. This often translates into either a perception of shifts in purpose or apparent disagreements on how it got started. In the history pages of its website, for instance, Greenpeace itself notes the irony of there being multiple plausible stories that describe its founding.*

Here I present the statements in reverse chronological order.

> **Greenpeace's mission** (December 16, 2014): We defend the natural world and promote peace by investigating, exposing, and confronting environmental abuse, championing environmentally responsible solutions, and advocating for the rights and well-being of all people.
>
> We take action only where we and our supporters have the most capacity to make an impact, where people's lives are most heavily affected, and where environmental risks are most dire.
>
> The problems we are trying to tackle are so big, that it takes a huge effort, comprised of people all over the world. That is where our supporters come in. Greenpeace is an inclusive, people-powered, collective movement. Our focus is on big political and corporate changes, just as much as it is on empowering people in our network to act in their homes and communities.
>
> **Greenpeace's mission** (December 1996): Greenpeace is an independent, campaigning organisation which uses non-violent, creative confrontation to expose global environmental

problems, and to force the solutions which are essential to a green and peaceful future.

Greenpeace's goal is to ensure the ability of the earth to nurture life in all its diversity. Therefore Greenpeace seeks to:

> Protect biodiversity in all its forms.
>
> Prevent pollution and abuse of the earth's ocean, land, air and fresh water.
>
> End all nuclear threats.
>
> Promote peace, global disarmament and non-violence.[*]

Greenpeace founder's recollection (from 1996 interview): We were just a handful of people from different backgrounds, but on one thing we agreed—this planet is in danger.

Robert Hunter's description of the mission of the founding trip to Amchitka (as printed in Rex Weyler's chronology): "The United States will begin to play a game of Russian roulette with a nuclear pistol pressed against the head of the world. ... There is a distinct danger ... that the tests might set in motion earthquakes and tidal waves which could sweep from one end of the Pacific to the other."

Billboards all over Vancouver (Green Panthers): "Ecology? Look it up! You're involved."

Description of the SPEC committee, circa 1969 (Although SPEC was technically separate from Greenpeace, several future members met through participation in that organization): The preservation and development of a quality environment through the stimulation of public interest, and consultation and cooperation with industry, government, labour, and academic communities.[**]

---

[*]Obtained from http://web.archive.org.

[**]Frank Zelko, *Make It a Green Peace! The Rise of Countercultural Environmentalism*, Oxford University Press, 2013, p. 38.

What may come as a surprise is that the process you need to follow before the problem and solution are clear follows the same scale-up logic. With the possible exception of highly regulated processes that require a fixed starting point, the evolution from what *The Innovator's Way*[8] calls the "sensing" phase to the "execution" phase can be smoother than most of us believe at first. No pivots are necessary. Placing the emphasis on scale and on making the problem tangible makes *your* process smoother.

Eventually, your activity will progress from discovering the problem to managing the organization that solves the problem. Certain words that seemed obscure at the outset—"team," "culture," "need," "product/service," "value," and so on—will acquire increasingly precise meanings that are given by the choices you make as you progress through the earlier phases of your inquiry. As you progress through the phases, you'll find yourself making more and more use of concepts from innovation management and finding entrepreneurial recipes you can use as routines.

Most innovations, as the Ford, Greenpeace, and even Kinect examples show, actually emerge while you're working on what you think is something else. There's no evidence that Henry Ford set out to create a new kind of assembly line. Johnny Chung Lee set out to emulate the movie *Minority Report*. That's why the notions of market pull and technology push, which paint formulaic pictures, actually conceal a richer reality in which any and many innovations are allowed to emerge from the process.

Learning as you go

If you are truly innovating, then what you are about to propose has not yet been done. Yet, as you start, you are mostly putting readily accessible parts together. As a matter of fact, you may not even fully understand the implications of your assembling until you see all those parts working together. This is one of those paralyzing paradoxes that, come to think of it, you *should* face as you engage in your inquiry. The only way out is to accept that you have learning to do as you go.

You are not an expert at the outset. You become an expert. It follows that, before you are correct about the problem, you are going to be wrong about it—essentially every single time. But you will be right about many smaller problems that you solve along the way.

As you collect all the elements of your innovating-as-problem-solving, to you it will probably look like this: many distinct, tangible problems besides the *one* that got you started; information obtained through conversation; several failed attempts; and a lot learned. The innovation may come from how you address some of those other problems. Henry Ford faced an inability to produce a sufficient number of cars per hour. He met that manufacturing challenge with "innovation"—the assembly line. Similarly, your innovation will come from things you had to conceive to close the gap between a tangible problem and its solution. That one problem that matches the hunch that got you started really emerges from an act of synthesis.

To others, viewing innovation in hindsight, your progress will only be apparent when they can imagine a product or a family of products. In other words, by the time you're nearly done, others will conflate your product and the innovation. As a good problem solver, you will revisit your problem and view your innovation in hindsight. Then you will be able to rationalize what you did into "stages" that may or may not have actually existed as you were going through them. You may identify phases of the work as gates and hand-offs and even persuade yourself that there was a specific moment at which you pivoted from ideation to execution. Again, though, these things are only part of your hindsight; they don't mean anything when you're innovating.

At the root of this process is something with which we are fundamentally uncomfortable. As we grow up, having acquired expertise in a field and tasted "being right" and being recognized for it, the very idea of being wrong is in itself traumatic. To accept being wrong as something good and useful requires a change of attitude, a shift in how we think about things.

Assuming you are not so lucky as to have learned everything you need to know before you even develop your hunch, at best you and the people around you have only *piecemeal* expertise and skill relevant to the problem. That makes it all the more sensible to accept that starting with parts lying around can lead you to something being recognized as an "innovation." That's the aspect of innovating that resembles problem solving. Along the way, people, skills, and clarity all accrue until impact can be measured. Learning helps you gauge progress.

I am often asked whether innovation (which represents a conclusion) can be learned. That's the wrong question. What matters is that *innovating* is something you can practice and become better at—if you achieve the right combination of knowledge and "muscle memory." The latter is possible only with perseverance. It requires abandoning recipes and engaging your brain. That happens to be what your brain is good at. Once you learn to trust your brain to operate outside the realm of formulas, innovating will become almost second nature.

Takeaways

• You can engage in innovating just as you would engage in problem solving. All you need to get started is a hunch. You can give any hunch the structure of a problem. Your job is to make the problem tangible.

• What makes a problem a problem is its structure. The facts that may ultimately inform the impact of a solution are just that—facts. What distinguishes a real-world problem from a fact is *recognition* (What are you trying to accomplish and why is that a problem?), *verification* (How will you know that you are done?), and imagining a *solution*. Your initial solution, your verification recipe, or even your problem itself can be wrong. That means you can get started from anything. A hunch will do.

• Innovations are not prescribed; they emerge from what you do in the process of trying to understand and tame a real-world problem, in the same way solutions emerge in problem solving. The word "innovation" is not shorthand for a new product. Innovations are not predictable at the outset. Acknowledging that all you have is a hunch may require accepting that everything is a variable, is subject to change, and that there is nothing to solve for—yet. That too is up to you to discover.

• Get ready to be wrong, and get ready to be told you are wrong even when you're not. A good solution can emerge from being wrong a lot. You are not an expert at your problem yet; expect to get to the "right" solution through a long sequence of "wrong moves."

• You need only be approximately right once. By then, you'd better have come up with your own way to know you're right: That is how you know you're close to the end. Otherwise, you may "innovate" past a solution.

• There are several advantages to prototyping a problem as an approach to innovation:

  Progress is about how much you learn about the problem.

There are multiple strategies for making your problem tangible and getting to specific questions.

There is a demonstration possible of any problem at a scale that matches your current resources.

• You can just get started by being wrong. This may require a shift in mind-set.

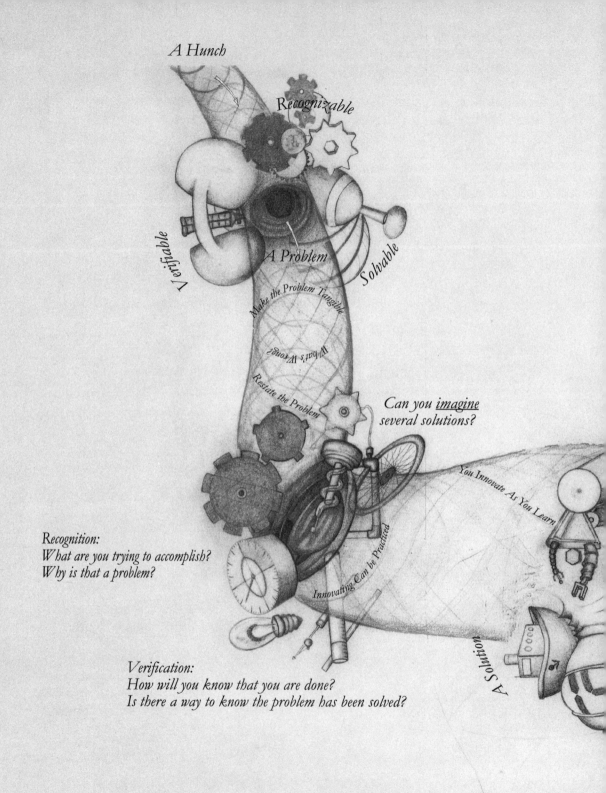

A Hunch

Recognizable

Verifiable

A Problem

Solvable

Make the Problem Tangible

What's Wrong?

Restate the Problem

Can you *imagine* several solutions?

You Innovate As You Learn

Recognition:
What are you trying to accomplish?
Why is that a problem?

Innovating Can be Practiced

Verification:
How will you know that you are done?
Is there a way to know the problem has been solved?

A Solution

PROTOTYPING PROBLEMS – INNOVATING AS PROBLEM SOLVING

At its beginning, no thing about an eventual innovation is new.
All you need is

> *A hunch about a real-world problem.*
> *A set of parts and a community to render the problem tangible.*
> *A strategy to engage in trial and error.*

Innovations are not prescribed.
The path – innovating – is full of choices and near misses.
Your hunch can lead to many innovations.
The impact can be social, technology, management, product, academic, and/or research.

A Tangible Demonstration of Impact

The Real Problem

You Solve

A "Recipe" to Know the Problem is Solved

II

EXPLORING IN FORESIGHT

Learning from Parts and People

IT TAKES LITTLE EFFORT to imagine an organization that serves a solution. Once you do, it is easy to divine that such an organization has some definite elements. Indeed, there is a constellation of definite elements—distribution, a channel, a business model, a product or a design, engineering, manufacturing, business development, regulations, users, advertising, purpose, impact, skills, and technologies old and new, to name some—that, with the help of people, you can conjure up to accomplish all you imagine. You imagine all this as some harmonious whole. It's all in your head.

It gets worse. The harmony implies that all these elements are interconnected in ways you designed—also in your head. But the only thing tethering this imaginary organization to the real world is your own set of experiences.

Those experiences might be all you need for your imagined organization to make sense to you. But it is still imaginary. You need to make it real. When you do, you'll discover that those elements live at the intersection of multiple disciplines, and that they affect one another in ways that are up to you to specify.

Netflix illustrates this well. There was a time when "distribution" implied a web page, your mailbox, the postal service, special envelopes, a logistics design, a network of DVD warehouses, rights to rent DVDs, and a subscription service limiting the number of DVDs you could have at any one time. It later implied a web page, a media player, a broadband connection, data centers, streaming rights, and an all-you-can-watch subscription plan.

So, it would seem that for Netflix the element "distribution" meant two very different things. Once Netflix began to produce its own content, "distribution" might have acquired yet a third meaning. Netflix could afford to have multiple meanings because the problems the company solved through distribution were real.

Back to your imagined organization. Sure enough, things eventually have to get to their intended beneficiaries. That, however, is not saying much about what makes a distribution strategy successful, or even what makes it real. Now consider the constellation of elements I mentioned earlier. Given the extent to which each piece of that constellation depends for its definition on every other piece, you've got to wonder whether any of the words that will at some point help you structure your organization even apply now. I do.

Before anything is defined, beyond divination, you've got to wonder whether you have anything other than the prospect of many moving parts and people that have to work harmoniously—in your head and outside your head—to be effective. And you have to consider all that implies in terms of scale. The organization you imagine is impactful already in your head; if it weren't, you might not even act on it. The real question—the real scale-up challenge—is *what* of that magical idea has a chance to escape your head and positively affect other people's lives.

The field of management has made it easier to phrase that question for business ideas. But the conundrums you face when making your imaginary organization real apply equally to sharing art, music, technology, or social causes. It is in your head, and before anyone else enjoys what's there, it needs to get out of there, become real, and somehow reach other people. It is only natural to expect that inside your head you haven't thought everything through in the same way the world might need you to.

Before we can even discuss the things that make management, large-scale engineering, design, marketing, and so on useful to you, you have to get what you imagine out of your head. For now, the least you can assume is that there are moving parts to your idea and that you'll encounter other people along the way.

That's my starting point for what I mean by parts, people, and how to interact and interface with them to attain scale and impact.

INTERACTING WITH PARTS

Innovating need not begin with raising lots of money. You can start with what you already have.

By now you should have a pretty good idea of a problem, based on your initial hunch. From chapter 2, you know your problem might still be wrong. The discerning reader will likely appreciate that there are always ways to be wrong for less money and fewer resources. You could ask for the money to address the big problem, or you could bring the problem down to a scale the available resources permit. Doing so lets you actually begin to work on solving your problem right away. You can start with parts lying around, as in the examples in chapter 1.

In other words, bring resources to the problem or bring your problem to the resources. It's your choice. Whether you postpone your start until you have all the resources you now think you need or you start right away, when you *do* get started you'll first have to figure out how your resources are going to work together. That's my starting point for the discussion here.

Resources are parts, people, funding, or knowledge. Funding comes from people, and knowledge may come from other people or from how you bring certain parts together. That leaves parts and people to discuss in this chapter and the next chapter.

Because a book requires some linearity, I had to choose whether to talk about people or about parts first. What I want to tell you about interfacing with people and interacting with parts are very similar, so I flipped a coin. Parts won, so this chapter focuses on parts; chapter 4 focuses on people.

I'll go more specifically into what I mean by "parts" later in the chapter. For now, you may benefit from approaching parts virtuously on the basis of this general principle: Whatever you dismiss as not being a part is something you prevent yourself from potentially using. You probably don't want to prejudice yourself in that way. If you think about parts in this broad manner it will probably aid you in the singular purpose of getting tangible as quickly as you possibly can.

> Start innovating with whatever you have. You renounce whatever you dismiss as not a part.

The fear of getting tangible

In my teaching, I have often encountered a fear of "getting tangible" too quickly—a fear about parts. This often translates into a belief that innovators-in-the-making don't really have the choice I mentioned at the beginning of this chapter: start by raising money or just get started. Aspiring innovators want to fall back on the recipe that would have them set out on a quest for funding with which to buy the expertise they think they need to take the next step. My experience and my observation of this phenomenon have convinced me that what they actually fear is their lack of knowledge. When that fear kicks in, they default to whatever will decrease their perception of uncertainty.

The winner tends to be recipes for pitching ideas or rushing a product out the door.

Fear gets in the way. You really don't need to know every single thing there is to know about a part before you use it. Actually, the initial purpose of making the problem tangible is to make clear to you what you know and do not know about your problem and what the current version of your problem requires of any solution. There is always at least one part you need that you already know how to use—like the billboard in the Greenpeace story. Later, in hindsight, you'll distinguish a sequence of fact-finding "missions" that took the form of experiments or conversations that revealed the nuances of the problem you seek to solve and the parts or people that led you to insights.

Parts and knowledge

In foresight, these experiments can be sorted by what you stand to learn from them and how much they cost. The guiding principle should be that you want to learn a vastly disproportionate, indeed *unreasonable*, amount with the money you have now, rather than postponing learning for when you have more money. That is how you choose the facts to seek first. I discuss this in chapter 9.

To be clear: the point here is not about bringing parts together to build the next gizmo. It is about acquiring knowledge about a problem from those parts—in much the same way as you acquire knowledge from other people (discussed in chapter 4). With the resources at hand, you bring parts together to get the most information about the ways in which you are currently wrong.

Don't postpone learning; learn a disproportionate amount with what you have now.

Parts and the Questions They Answer

In essence, "parts" are everything that are not people. *Everything.* If your idea evolves into a commercial organization, for instance, the actual components of a gizmo, your service offerings, the distribution channels, and even components of a strategy will all be parts that survived your innovating. You can give them names at this early stage, but I recommend you simply think of them as interchangeable building blocks—parts—that you are going to play around with a lot. You'll dispose of some, and you'll enlist some new ones.

How these building blocks interact with one another is your choice. Their function will change as well. At first, they will mostly serve the purpose of making the problem tangible for you. Later, they will have to make the problem tangible and solved for others as well.

From an operational standpoint, a part is anything you can order, procure, simulate, or otherwise use to make your problem tangible. To identify the parts you need, you may have to demand more from your hunch and expect it to answer questions like these:

- What makes this hunch so special? What do I need to obtain to simulate that?

- What could go wrong? What do I need to do (experiment or otherwise) to show that it is wrong or assess whether it is even possible?

- Can I build something with which to demonstrate the function my hunch calls for?

- Is there an experiment that will give me more information or demonstrate a specific feature my hunch may be lacking?

In one of my classes, a group of students wanted to consider various opportunities related to the so-called smart electric grid. They needed an electrical grid to play with. One option was to start coding to create a software simulation of an electrical grid. The problem with that is that one such simulation would be only as good as the assumptions they made in coding the software.

Fortunately, nature computes in real time. So, rather than a software simulation, I suggested they think about a version of the problem that could be set up at table scale and then begin by simulating what the solution would have to accomplish at that scale.

In the end, they needed only a few cables, some variable generation sources, and a means to simulate demand patterns with a few variable resistors that they could coordinate with microcontrollers. Nothing they used was any more complex or expensive than stuff you could pick up for a few dollars at your local hobby shop.

I contend that you can acquire knowledge you need about your problem as you try using parts, and my experience is that just about anyone can be trained in a single day to use a wide array of parts purposefully for innovation prototyping. It all boils down to two things. One is looking at parts in a different way as meaningful, tangible representations of aspects of your problem. The other is accepting that you don't need to know everything about what makes a specific part work; you need only learn how to use it in the context of your problem.

This is not an argument against whatever knowledge you may already have—which is valuable—or an argument in favor of outsourcing your innovating. It is an invitation to view that knowledge as an asset rather than allowing it to become a constraint—an idea discussed later in the chapter in the context of what it means to use accessible parts and knowledge.

You really don't need to know every single thing there is to know about a part before you use it.

I can give you a simple illustration of how, with minimal instruction, you can learn to use parts and build from them without any previous knowledge of why those parts work together. Instructions for creating rainbow-colored milk are offered on the Internet as a STEM activity to do with kids; the parts you need are whole milk, food coloring, a dish, a cotton swab, and soap.

For older children—potential young scientists—the activity is presented as way to teach about chemistry, the concept of hydrophobic and hydrophilic groups, and repulsive forces. But the outcome is the same regardless of how much chemistry you aspire to learn: When a cotton swab dipped in soap touches the thin layer of milk in the dish, colors rush away from the drops of food coloring and begin flying around. Toddlers are thrilled by the visual experiment and enjoy playing around with the parts. Curious children, regardless of age, will want to do variations on the experiment, try different liquids, or build on the experiment by trying something new, such as creating a video of the experiment to show at school. Knowledge of chemistry, though handy, is not required in either case. Intuitive knowledge of chemistry, however, can be an outcome.

> You can acquire knowledge you need about your problem as you try using parts.

After following the instructions a couple of times, you may wonder what else you might do. That's when the fun begins.

Perhaps you want to understand why you see the colors rush away. That is precisely what the instructions will entice you to do—act like a scientist, they say. You may change the substrate from whole milk to something else. You may measure how long it takes for colors to fly about. There are many things you could try. You'll learn to master the experiment.

Perhaps you don't feel like acting like a scientist today. You may wonder—now that you know how to put the parts together—what else you might use this activity for. For instance, you could modify the shape of the plate and place colors strategically so that when the colors start flying around they spell out a name. Maybe you'll record your experiment and post it on YouTube.

There are subtle differences between these two directions. In the first, your objective is to characterize a phenomenon. The second is a way to "use" science—regardless of how much you understand of it—for a purpose other than learning the particular science itself. It's a lot like "working" with your hunch when it is not yet fully formed: you may need to play around with some parts to make the purpose clear to you.

Choosing to "use" the science doesn't mean you waive understanding the underlying science altogether. Indeed, as you perfect your idea, you will still find yourself needing to experiment further, and you may end up doing some of the same experiments your "scientist self" would have done. Your purpose, however, will be very different: Rather than recording the chemical and physical behavior against a set of abstract parameters, you will focus on building something you envision. You will end up picking up part of the science intuitively or by asking questions of experts (who are people, of course, and so are the subject of chapter 4), but the knowledge you acquire and the questions you pose will be driven by what *you* want to accomplish—not by the general model of science the instructions of the original experiment want you to learn. As a rule of thumb, when your purpose isn't to understand the science itself, the knowledge about the science you need can be more targeted and you can acquire it as you go. Either way, the parts are the same. What changes is how you go about using them.

Parts for "use" and scale

In this chapter, I ask that you think of parts as in the second direction above, driven by what you want to use them for. In a way, I am asking you to take on faith that whenever you need to understand more fully why parts work, you'll be able to do so by "debugging" or by asking targeted questions of others who you believe to be more knowledgeable than you are about the underlying phenomena. Of course, you won't really know whether they are until you ask *your* specific questions.

The example above is but a cursory introduction to how to think about parts. All the parts are material, and the science is accessible. However, everything I have said about the two directions in the rainbow-colored-milk example applies to more complex science and also to social and organizational measures you might propose to address a problem. You can indeed produce parts, workflows, and dynamics with which to simulate, at scale, just about anything.

Implicit in this mind-set is that the focus on a real-world problem invites you to imagine that every part you bring into your problem is, in a way, a scaled-down representation of a larger reality. With a sense of what it may take to solve your problem, your objective is to "demonstrate"—first to yourself—what makes the problem real at the smaller scale that matches your resources. That begins with parts.

"Parts" defined

What do I mean by "parts"? I am using the word as generally as you can possibly imagine. I want to disabuse you of any sort of binary thinking you may employ when it comes to parts, as in "this is a part" and "this is not a part." A part does not have to be something like what your auto mechanic would get from the "parts store."

The simple definition of "parts" for innovating is this: whatever—and I genuinely mean *whatever*—you can bring together at a small enough scale that illustrates something about the problem and, in doing so, reveals what you may want to do next. In other words, parts in this context are what help you specify your problem, the result of which is one or more versions of a better problem to solve.

Conversely, whatever you exclude from potentially being a part might have helped you—but never will. So, rather than pondering whether something is or isn't a part, you may want just to assume it *is* a part and instead ask what you might be able to do or demonstrate with it.

If your parts are diagrams, you'll illustrate an abstract concept, perhaps a business model. If they are plaster and cardboard, you'll illustrate form. If they are mechanical gears and microcontrollers, you'll illustrate function. If they are org charts, you'll illustrate organizations. If they are patents and academic papers—yes, they can be considered parts—you'll illustrate the use of some basic science. As you demonstrate a problem, you may need to consider each and every one of these at some time. For the most abstract parts (such as business models or patents), you'll have to think about whether showing—say—a diagram is enough or whether you are better off somehow enacting what it shows.

One way to get started thinking about parts in this manner is to work backward from your problem. You may think of parts as elements of your imagined solution. Or you may think of parts that help you reproduce something your imagined solution must achieve. Or you may think of parts that

Parts are *whatever* you can use to illustrate something about a problem and reveal what to do at the next scale. Parts represent a larger reality.

help you verify that a proposed solution actually solves your problem. These are the same three structural elements of well-defined problem. Any of these parts might be part of an artifact, part of a larger system, or an auxiliary part whose main purpose is to inform the next version of the problem you're solving.

Auxiliary parts deserve special attention. When you are learning to draw, you are taught to begin drawing circles, geometric figures, and auxiliary lines to get the basic shape right. Those "parts" will not make it into the final drawing, but they do play a critical role in getting the shape and the proportions right. Similarly, the cranes and scaffolds used in erecting new buildings play a critical role in the process but don't make it into the finished structure.

The concept is used also in mathematics, in which geometrical constructions or lemmas simplify and often make it altogether possible to come up with a proof. On occasion, significant innovation goes into developing these "auxiliary" parts.

Ford's assembly line could be considered an example of an auxiliary part that became an innovation. In the rainbow-colored-milk example, the color is an auxiliary part. The phenomenon you observe is the same one you get when you use soap to clean a pan: Fat and oil flee away from the soap. If you are using the activity to learn science, food coloring is but an auxiliary part that aids visualization. If you opted for producing a video, the food coloring that was an auxiliary part may become the most important part in your experiment, and you may find yourself experimenting about the effect of color.

Choosing parts by the aspect of your problem they make real (based on examples found throughout this book)

Type of part	Imagined solution	What needs to be accomplished	How you verify your solution
Component of an artifact	Flash lamp from photography catalogue (laser)	Raspberry Pi (CreepyDOL)	Altimeter (X-Prize)
Element of a larger system	Billboards or ham radio (Greenpeace) Business model	Conference (CreepyDOL)	Performance standards, such as those that might be established through regulation (e.g., E911 standards)
Auxiliary	Strawberry (infectious diseases)	Power supply (physical simulation of Smart Grid)	Focus groups

The purpose of parts and prototyping

You need parts to prototype each and every aspect of your problem—from the reasonably obvious components of a gizmo to the less obvious aspects of the organization that will serve the solution.

I am using the word "prototyping" in a way that may seem atypical. Typically, prototyping is thought of as a means to create an artifact or a version of a solution, and a part is but one element of that solution. You imagine your solution will be some sort of phone, and so you set out from the beginning to include a part that is a screen. But that is just one of the ways of thinking about parts I've just given you.

The way I have introduced prototyping a problem in previous chapters gives you a choice as to what you start assembling. Sure, you may assemble a mockup of a solution, but you may also try to bring parts together to simulate how the world would look once the problem is no longer there, or you may choose to simulate the problem itself or assemble parts that will allow you to verify that the problem is solved.

Your choice of what to prototype may determine the parts you need to acquire. Conversely, the parts you have available may make it easier to prototype something other than the solution.

On non-tangible aspects and parts

Thus far, my examples have involved mostly material parts. It would be only natural for you to object that this approach pertains only to problems that can be addressed with some kind of a tangible gizmo. Parts, though, are but a means to reproduce function at scale, and you can reproduce function of tangible *and* non-tangible aspects of your problem alike. For instance, some aspects of your problem can be and often are less tangible than others, and their function may be best introduced into your innovation prototype with workflows and processes that affect how the entire organization you build functions at scale.

Here is a straightforward example (like the one above for material parts) in which the non-tangible aspects are a mixture of a business model and an analysis of a supply chain—both parts: You probably buy your apples from a grocery store that gets them from a distributor, or from a farmers' market to which the grower brings them directly. Before that, the apples were picked, either by the grower or a crew of workers, and before that the grower tended the trees. You pay the grocer, but you can think of your money as being divided into different portions allotted to the grocer, the grower, the pickers, the distributors, and so on—although in reality all of that is largely settled before you buy your apples.

Where I live in New England, many families go apple picking as an outdoor activity in the early fall. In that context, the same orchard, the same grower, and the same trees now address a different problem altogether and have become part of that family activity. The difference is two non-material parts that you may not typically think of as parts: the business model and the pickers.

Whatever problem you envision has both tangible and intangible aspects. By the time the problem is solved, you'll be able to point to a definite set of components in some organization that contributes to addressing every aspect of the problem, regardless of whether you initially thought of that aspect of the problem as tangible. Being able to point to those components now affords you clarity that is worth having as early as possible, if for no other reason than it will make it easier to change them if they don't work.

Getting this clarity early on is much easier than you may think. It requires that you adopt a principle: "Everything can be demonstrated tangibly in some way." Your job is to find a way to bring parts together so that even the non-tangible aspect of your problem can be illustrated tangibly.

The box titled "Even Non-Tangible Aspects of Your Problem Have Material Parts" shows how you may go about formulating a non-tangible aspect of your innovating in terms of material parts.

Even Non-Tangible Aspects of Your Problem
Have Material Parts

It's a challenge to think through the non-tangible aspects of an emerging problem or organization as something that can be represented with parts. Tangible aspects of a problem have technical dimensions and units by which they can be specified. They exemplify the specificity with which you should strive to understand even the non-tangible aspects. But those non-tangible aspects that may at some point become tangible—in the form of a business model, manufacturing in large volumes, distribution, referrals, a value chain, complementary assets, messaging, a value proposition, social impact, and so on—do not always come with the extra help of being pre-specified.

Still, your success could well depend on making the non-tangible aspects just as concrete and well specified—and there's a rather simple way to do that. It begins with the realization that what's tangible often changes with scale. At some point, those aspects will be tangible, too. There's got to be some way to simulate, illustrate, or demonstrate tangibly their function today. That is, you must dissociate what needs to be achieved from the nuances that will be introduced through scale-up.

The process goes more or less as follows:

• *Diagram.* If it will eventually be material and tangible, sketch an illustration of what will be achieved and how. Flow diagrams work well for this purpose.

• *Enact.* Have multiple people participate in enacting your flow chart to develop a "variational" understanding of what happens when things do not go exactly as planned. (If you've ever played *Monopoly*, you've done this before, enacting the life of a real-estate mogul at scale. Your "variational" understanding could come from erecting hotels on a property than no one ever ends up landing on. An enactment like that is a better demonstration of what ought to be accomplished than simply producing a flow chart.)

• *Emulate.* You can often come up with simple parts with which to illustrate, simulate, or emulate what you learned from your enactment.

If you have worked the non-tangible aspects of your problem into your innovation prototype, it will be more difficult for you to ignore them. It will be easier to reason about scale-up if they are a part of your innovation prototype from the get-go.

There are several tools you can use for inspiration, if not as parts, to help materialize non-tangible aspects of your innovating. The Beer Game teaches you how to enact and think about disruption in a supply chain.[*] Online A/B testing can show you how online users interact with your idea.[**] The business model canvas, though generally limited in scope to consumer products, offers a "LEGO-like" view of common components of a business model.[***] I offer these examples without endorsement; choosing the parts that apply to your innovating is part of your job.

The following straightforward questions illustrate how you may go about turning the less tangible aspects of your innovating into parts:

- How will my idea get where it needs to go?

- What will people see as a demonstration that it works?

- What does a transaction look like? What else might people need in order to benefit?

- Is there a way to rearrange parts that changes costs, prices, regulation, and value?

[*]Jay W. Forrester, "Industrial dynamics: A major breakthrough for decision makers," *Harvard Business Review* 36, no. 4, 1958: 37–66.

[**]Ron Kohavi and Roger Longbotham, "Online controlled experiments and A/B tests," in *Encyclopedia of Machine Learning*, Springer, 2011; Brian Christian, "The A/B Test: Inside the technology that's changing the rules of business," *Wired*, April 25, 2012.

[***]Alexander Osterwalder and Yves Pigneur, Business Model Generation: A Handbook for Visionaries, Game Changers, and Challengers, Wiley, 2010.

Parts in action: a demonstration from my workshop

I contend that with your problem and some parts, you can take some amazing steps along the path to innovation. Allow me to tell you about a workshop I ran at MIT that demonstrated this very thing. It shows not only how powerful working with parts can be, but also how motivating it can be to engage in innovation prototyping once you realize you can dismiss the fear of failure in favor of learning.

A few years ago, forty-plus students walked into an MIT classroom to engage in the workshop. Most had come from abroad and expected to have plenty of time to take in the sights of the Boston area and even make a side trip to New York City. They were mostly wrong about the side trips—and not because I was a cruel taskmaster.

Twenty days and nearly 300 hours of mostly self-driven work later, workshop participants had produced innovation prototypes comprised of gizmos, websites, and primary market research. They had advocated for their innovation prototypes to an audience of people from MIT's vaunted innovation ecosystem, and they were ready to begin to address scale with clarity on what to do next to further their understanding of the problem.

Few, if any, visited New York City. Instead, they chose to spend their nights and weekends practicing a different kind of tourism: visiting stakeholders, learning about impact, and tinkering with technology. On the last day of the workshop, they presented their plans. One group demonstrated a system to manage inventories with machine vision—video of an on-site demonstration with a user and launch strategy included. Another showed a physical prototype of a smart electric grid that allowed them to formulate an academia-industry research project. A third group demonstrated a robotic system to increase patient compliance.

Three months after the workshop, I ran into one of the participants—a theoretical physicist—on the MIT campus. Before the workshop had begun, he had argued that the closer to a published paper, the purer the innovation, and that everything needed to be new to call something an innovation. Now, however, he was quoting chapters 1 and 2: "You have an idea? Okay. What should you ask? Who should you ask? What's the closest thing to your idea you can put together to start to make it real? What parts should you start putting together? For a handful of dollars, you can put a few technology parts together and become much smarter about which questions you want to ask." Then he opened his backpack and showed me an assortment of parts he was carrying around with him. He intended to prototype a real-world problem rather than study it mathematically and write a paper about it.

I led another workshop the next summer. This time around, we added options for participants to prototype biotech innovations—something many thought could only be done with access to specialized labs. The workshop results were just as impressive. We got to see plans and prototypes for an organization working to shorten the time it takes to diagnose infectious diseases with a portable DNA analyzer (mentioned in chapter 2), for robotic irrigation systems, and for a machine-vision approach to turning social networks into real social support networks.

This happened because these participants had done what the theoretical physicist had quoted from me (that appears in chapters 1 and 2). Workshop participants learned to acquire knowledge on-demand—when they needed it— rather than encyclopedically. They reasoned about problems to be solved, and learned to own their projects through quick and productive failure events. They came to understand the difference between decidable problems and causes, and practiced how to refine problems by tinkering with technology and impact together.

They also learned that the first steps in innovating are fueled more by bringing together new combinations of parts than by raising large sums of investment capital. Some call the outcome of this combining of parts "creativity." The parts—*whatever* parts you bring together—play a big role in conveying that you don't need lots of money to start, which workshop participants absorbed as they experienced the artificial constraint on the resources to which they had access. You can find what you need for prototyping pretty much anything simply by searching for inexpensive and readily

A Discipline

Many Disciplines

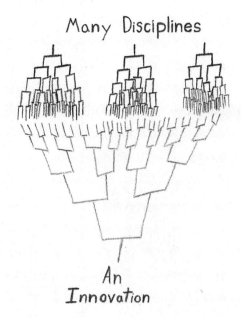

An
Innovation

An innovation is generally supported by knowledge from multiple disciplines, but only a fraction of that knowledge. It might be more advisable at times to think of an innovation as a discipline of its own and acquire knowledge on demand, just as other disciplines acquire knowledge.

available parts online, and get them in only a day or two. Dr. Maiman explained this very principle when he described how he developed the first working laser, and Steve Wozniak alluded to it when he described what he thought he and others were doing when the Apple I came to be (see chapter 1).

A World of Available Parts

There are parts that you already own. There are myriad technical components and parts you can order online. Beyond that, there are educational resources, including complete course outlines, books, MOOCs, and so on, all easily accessible and all also parts. Your need for any particular part may be immediate or become apparent only down the road. The latter is probably the case with online services that help you build websites painlessly, or marketing and strategy resources that include toolkits for generating business models, doing web analytics, and conducting A/B testing (randomized experiments with two variants)—also, all parts. Then there are outsourcing resources that run the gamut from accounting to contract management to order management and fulfillment, not to mention small-batch contract manufacturing. Again, these are *all* parts.

At table scale, though, it is an abundance of inexpensive parts that aid you in prototyping virtually anything. You are the beneficiary of a surge of resources made available by do-it-yourself enthusiasts and open-source communities, and by the trend toward "democratizing" innovation. The Web has become a paradise for hands-on-innovators.

Where do you begin your search for parts? The answer comes from asking yourself these questions:

- What do I need my innovation prototype to *do*?

- What is available to me that gets me close to that?

- What do I want to show as possible (first to yourself, and eventually to others)? What do I need to show?

- What can I acquire for that purpose at a cost that doesn't make me think twice?

Only you can decide what is an "affordable" investment in your problem. In the workshops I've run, we set the limit arbitrarily at $100 per part. To get a part that costs more, you have to explain why you would be willing to pay for it yourself.

A part is as good to you as the function or the aspect of your problem it allows you to demonstrate—no matter how unsophisticated the part. The $100 limit is an artificial way to force people to question how invested they really are. If you aren't willing to cough up a few hundred dollars to make your problem tangible, then you are the first one who doesn't care enough about your problem. You shouldn't find it surprising that others don't care either.

The specific parts you use at any time are important because they help you understand your problem, not because they may or may not be part of an eventual "innovation."

The best thing about working with parts as I've described is that it is liberating. The early prototypes in the stories I mention in chapter 1 are indistinguishable from what may have been the starting point of any "failed" company. That is, the innovation is not in those early prototypes any more than it needs to be in yours. It is in what is yet to come. And the value is not in the prototypes per se but in what they demonstrate to be possible. You get to go about innovating the way Dr. Maiman acquired parts for his laser, possibly with greater ease. He had to read through catalogs; you get to search for your parts on the Web.

Working with parts at scale

A broad definition of parts is not only proper for innovating, but also makes your life as an innovator easier: It removes unnecessary constraints on how you think about innovating.

To you, parts come in two flavors: parts that help you represent a tangible aspect of your problem and parts that help you emulate something that you will only understand as tangible at a larger scale. If your part is material, you'll have to figure out how to put it together with the other parts and then set it in motion, turn it on, or point to it. If it is a diagram, you'll have to decide between showing it and enacting it. The more physical the parts, the easier it is to converse with them and have them tell you what makes your idea wrong.

The parts you use help you understand your problem even if they never become part of an eventual "innovation."

Essentially, your task is to find a part for every aspect of your problem you want to make tangible at the current scale. To get to the next scale, you should strive to make as much of the problem tangible as is needed to persuade you it's worth spending any more of your time on it. You may think of yourself as "investor zero." Getting to the next scale will require letting your imagination run a bit wild and pretending that your parts represent something else. This can be fun to do. Chapter 2 includes one such example: the team of students who pretended a strawberry could be a bacterium so they could imagine a solution to increase successful treating of infectious diseases by expediting diagnoses. Using the strawberry allowed them to postpone costly access to a facility with a biosafety level of at least 3 at no risk to them or others, all the while learning what needed to be accomplished once they sought that access.

I got that team started with a simple exercise not much different from making rainbow milk. Simple and easily obtainable parts were required: one strawberry, one resealable plastic bag, one teaspoon of salt, two teaspoons of

Would You Have Ever Thought a Strawberry Could Be a Bacterium?

Probably not, but a question like that only emerges in hindsight. Here's how you get to an answer that motivates such a question.

You could spend a lot of money to rent an adequate facility, hire professionals skilled at sample preparation, procure bacteria with all the regulation and safety considerations that entails, plan experiments, and so on—all to assess whether you might have a way to identify bacteria rapidly through DNA analysis.

Might there be a faster and less costly way to try to prove yourself wrong? After all, everything hinges on the risks involved in handling one particular part: bacteria. Is there a way to replace that part?

If your thinking got you to the point of asking that question, you might turn to a friend who knows more biology than you do. That's what I did, and I learned that a strawberry contains eight strands of DNA—four times as many as you do—which makes DNA extraction, preparation, and detection easy. It's a strawberry, so unless you're allergic you know it's safe.

It turns out that for the purpose of testing an idea, a strawberry is indistinguishable from a bacterium—and a lot easier to handle. Plus, you can conduct a scaled-down version of your experiment in a kitchen, a place for which you already pay rent or perhaps even own.

To be sure, this will change the order in which you do some things. You'll still need access to specialized facilities—eventually. But by then it will be to answer highly specialized, targeted questions that justify the expense.

Incidentally, those are the targeted questions you arrive at by asking yourself what differences between a bacterium and a strawberry matter for whatever you're doing. Those questions are in large part associated with scale. Once you have a strawberry-powered prototype, it becomes easy to address tangibly what those questions ought to be.

dish detergent, half a cup of water, two plastic cups, one coffee filter, 90 percent rubbing alcohol (ice cold), paper towels, and some coffee. As they prototyped the problem, they added other parts: some tubing, a few small model-size motors, some LEGO blocks, a microcontroller no more sophisticated than the one in your house's thermostat, and eventually a PCR chip, a portable electrophoresis device, a strategy for deploying their idea in developing countries, and a website explaining the project and the organization. In the end, they demonstrated—at scale—a not-for-profit organization to diagnose infectious diseases in developing countries an order of magnitude faster than is typically done today.

When you work with parts the way I propose, you can view the scale at which you demonstrate your problem and your solution, or certain aspects of your problem, as a choice.

Also in chapter 2, I describe how some aspiring innovators wanted to prototype a pill that would signal when it had dissolved fully. The problem they wanted to solve was patient compliance. Unfortunately, they fell in love with their product idea and ended up obsessing so much about the impossibility of prototyping a small enough pill to demonstrate the concept that being right about that impossibility became almost more important than the pill itself.

Regardless of the merit, there was a way to prototype their pill idea and get to the questions that would have allowed them to improve what they wanted to accomplish. Instead of obsessing about the small size of a pill, they could have treated the human body as if it were that of a giant. Then, to prototype their idea, they could have used accessible, affordable parts—for example, standard-size, easily accessible RFID chips; cereal to coat the chips; some kind of container to simulate a larger-than-life human stomach; and some fat to coat the container (for a realistic representation of the giant human in the prototype).

After that demonstration, they could focus their "obsession" towards a more productive endeavor: the specifics involved in scaling up—which, in this case, implied also bringing the pill artifact down to realistic human size. Is this solution meaningfully addressing medication compliance? What parts are missing? How strong a signal does the swallowed pill need to send?

What is required to pass regulation (a part)? What would human testing involve (a part)? These questions all point to the next set of parts needed for innovating, and they drive the conversation away from the initial obsession about a pill and into specifics about where they might need to innovate.

Sorting Through Parts to Identify Non-Tangible Aspects of Your Problem

The material parts you may use to imagine the larger reality are easier to identify than the parts you may use to emulate non-tangible aspects your organization you will have to address. That is, it is easier to come up with an idea for a "gizmo" than to come up with a full system of organizations, regulation, services, and agents. The RFID pill is one such example.

Sometimes regulations and markets stipulate specific standards to which you must adhere; you can think of these as parts and reproduce them as checklists. These checklists are easier to follow against a prototype than they are in abstract.

Finding ways to make organizational parts tangible is a key part of your innovating. For next to nothing, you may find resources online that can help you emulate or prototype an organization. You can find legal information and templates for a variety of needs online. The websites of government agencies guide you through what's required, and government regulations offer specifications that may guide you to select certain parts over others.

But there was no guarantee that the pill the students had imagined would solve the problem of patient compliance they had set out to address. Perhaps, though, what they learned and the specific questions their first prototype triggered would lead them to restate their problem in myriad new ways. Before I had suggested a path to bring the problem to table scale, they had argued that they had no choice but to raise money to produce a proof of concept. There certainly are ways to be wrong more quickly with less effort and less money—and to learn something along the way.

Parts demonstrate a larger reality

At first, coming up with parts for imagining a larger reality can feel some-what daunting. But all you really have to do is put together a demonstration of your problem—a demonstration that is not only visual but also functional and tangible, with parts you can move around. If all you have are Post-its, you can most certainly start there and outline each and every element of the problem and how you will solve it—so long as you can resist the urge to "brainstorm" with those Post-its. But Post-its can't demonstrate function, so sooner or later you'll have to come up with a way to replace some of them with useful parts that *do* something.

Example of Specification of a Non-Tangible Part

Let me take you through an example of how you may specify parts that illustrate non-tangible aspects of your problem in a way that allows you to imagine multiple larger realities.

As chapter 4 shows, you are always better off thinking about your innovation as eventually serving a community, mainly because members of a community speak to one another. If they did not speak to one another, you would have to find a way to speak to every community member individually, which would make for a rather poor scale-up strategy.

You need a part that captures how the word will spread in a community. Let's call it a part for *referrals*.

There are all sorts of referrals. If your innovating leads you to a mass-market retail product, your best referrals may come by attracting a celebrity who will endorse what you've created. You'll need to know how much the celebrity charges to be the public face of your product, and you'll need to think about the community the celebrity will reach.

If you are thinking about an industrial product or a commodity chem-ical, your referrals will come from your existing customers. You will need to think of each customer as an investment in a demonstrable outcome and in terms of the communities they may be able to reach. Note that in some cases the communities may be in industries that are somehow adjacent.

You can specify the role of the non-tangible part *referral* in your innovation by at least two parameters: the investment required to create a referral and the kind of community a referral will unveil. You can use referrals as a part in your prototype to represent all three larger realities above. You may eventually call these referrers lead users, strategic partners, key opinion leaders, early adopters, or something else. However, the fundamental characteristic that makes them attractive to you remains the same: Investing in referrals stands to open up a community.

Another quick example is implementing the "affordable" aspect of the Model T. Ford ended up requiring at least three non-tangible parts: a financing model, an increase in wages, and a method to assemble cars efficiently to increase daily factory output. These non-tangible parts enabled and constrained different aspects of the tangible parts no less than the engineering design of the car.

Once you get used to this way of thinking, you'll find it increasingly easier to see parts for their demonstration value rather than for what they "officially" are meant to do. Think of it a bit like pretend-play, not unlike how, at some point earlier in your life, you may have found it perfectly natural to see a sofa for the castle it conceals—two throw pillows away from becoming a spaceship. Let me get you started with some examples of how you might use parts to imagine larger realities.

Parts mind-set: seeing a sofa for the castle it conceals—two throw pillows away from becoming a spaceship.

All Kinds of Parts

Here are some examples (an exhaustive list would be too long) of the kinds of accessible parts you may find across very different domains and that you'd likely be able to afford out of pocket (i.e., for about $100, on average). For illustration purposes, I emphasize things you may not have imagined are so easily obtained.

From the "technology" world: nanoparticles, microcontrollers, microfluidics kits, PCR machines, all kinds of sensors (and motors and lasers), mini-computers, portable mass spectrometers, DNA primers, strawberries, backyard wind turbines, scientific and educational kits, software-development kits (Apple, Bitcoin), product-development kits for anything from robots (LEGO, iRobot) to wireless electricity (WiTricity).

From the management world: marketing resources and components made available by the Lean movement, the Disciplined Entrepreneurship "toolbox" and its 24 steps, design thinking, permission marketing, inbound marketing, and so on; resources to view strategy and business models as composed of modules, including the Business Model Canvas and the Blue Ocean Strategy; online metrics and measures with which to assess impact, including A/B testing and online surveys; impact visualization tools such as D3.js, Google Analytics, Cytoscape, and so on.

From the communications world: painless development environments such as Wix and Squarespace; social media campaigns through Twitter, Facebook, and so on; crowdfunding.

From the legal world: regulations (governmental or market) and the specific standards they stipulate; legal information and templates for a variety of needs, such as those offered by LegalZoom; guides to requirements on government websites.

Finally, there is *knowledge.* You can go as deep as searching papers and reading the specification of patents, or you can get started with less. For nearly everything you may imagine someone can teach, you can find online a specialized educational or scientific kit that will make you smarter. At the very least, it will teach you what knowledge you need to acquire on your own or with others. For example, MIT's Nanomaker[*] will teach you how to build your own solar cell with sunscreen and raspberry jam. Spectruino and NZnano[**] teach how to build homemade spectrometers for under $500. There are all kinds of other online resources for acquiring knowledge through doing: instructables, the open-source communities surrounding Arduino and Raspberry Pi, online physics chemistry education resources,*** maker communities, and YouTube channels such Veritasium,**** to name only a few.

The bottom line: these examples make it nearly impossible for you to use lack of availability as an excuse for not getting started.

*http://ocw.mit.edu/courses/electrical-engineering-and-computer-science /6-s079-nanomaker-spring-2013/

**http://nznano.blogspot.com.au/2011/12/homemade-spectrometerspectrophotometer. html; http://myspectral.com/

***See, for instance, "Science Is Fun in the Lab of Shakhashir," at http://scifun.chem. wisc.edu.

****http://youtube.com/user/1veritasium

Making the most of parts you *do* have

You are allowed to be wrong. It's easier to identify that you are wrong and figure out why than to obsess over whether you might be right. What parts should you get, then? As you are getting started, just about anything that is accessible to you will do. You don't need any kind of technical education (whether in marketing or in engineering) to choose your parts. And whereas many years ago it might have seemed necessary to have access to a scrap yard, today you can order almost anything online, have it delivered to your door within days, and find basic how-to information on the Web. If you feel strongly enough about your hunch to derive some purpose from it, neither parts, nor people, nor knowledge are the limiting factor to getting started.

The following principles, also summarized in the next box, can guide you as you try to identify parts with which to make your problem tangible.

Principles to Guide Identifying Parts for Making Your Problem Tangible

1. Everything can be demonstrated tangibly in some way. The only limits to what kinds of parts you may bring together are limits you impose yourself.

2. At first, everything you might obtain is an auxiliary part: Parts extend your power over nature so you can demonstrate a problem tangibly.

3. Everything already in your possession is accessible and hence could be used as parts.

4. What you don't already own but need is probably easy to obtain at little or no cost. The alternative is to accept that what you need is beyond your resources.

5. Parts help you glean what might go wrong at the next scale.

Everything can be demonstrated tangibly in some way. The only limits to what kinds of parts you may bring together are limits you impose yourself. To develop the kind of seemingly preposterous propositions that led to innovations—Ford, Greenpeace, and other examples already mentioned—that we now consider obvious, you must allow for the possibility of bringing together parts that do not seem to be made to work with each other and ask yourself this: "If these parts *did* indeed work together, what problem would they illustrate?"

Among the wide array of parts you might use, some may help you emulate an organization, some may help you give form to your design, and still others may be "technological." If you do not think of yourself as a technologist, an artist, or a manager, you may have to overcome some fear. But the good news is that you don't need to know anything technical about the part; you need only use it—just as you don't have to know how your microwave oven works to warm up a meal.

At first, you may just follow some instructions. As you flesh out your problem, it will become clearer whether you need to modify or develop new technologies, what they ought to do, and where to search for any expertise you might need.

At first, everything you may obtain is an auxiliary part: Parts extend your power over nature so you can demonstrate a problem tangibly. This ought to help you overcome the self-awareness that comes from not knowing whether your idea is the *one*. That is, the purpose of parts is not to shape a product, but to help you think about the problem tangibly. There is no need to worry at this stage whether your final "gizmo" will include any variant of the parts you choose early on. That's true even if you came up with ideas for parts while thinking about a product.

This principle allows you to focus your search on what you have or what you might easily obtain. It also allows you to formulate follow-on questions about every part: "Now that I know how to do this, what else might I use this for? What else might I be able to use this with?"

Parts are meant for your own demonstration purposes. No need to worry just yet about what might go into the solution. The parts you choose will either extend your power over nature (that's how William Barton Rogers defined technology) or help you simulate something about how the problem you imagine connects with what actual people do (i.e., how it extends their power over nature). Either you'll have these parts already or you'll have to acquire them. The table below gives you some ideas.

It's easier to presume you're wrong and figure out why than agonize over being right.

How parts extend your power over nature

	Exists	Can be emulated
You already have it ("**What can I use that** *for*?")	Whatever you or your organization currently owns or routinely produces or procures for a different purpose.	Something tangible with which to emulate something else that is beyond your current resources; or a task such as a workflow or dynamic already used by some department or group within your organization.
You need to acquire it ("**What do I need to do** *that*?"	Something you can procure at a price that doesn't make you think twice.	A design or model that explains how you would go about creating a workflow, dynamic, or aspect of an organization (e.g., certain kinds of legal documents, a "dry-run" process, the specification of a type of test required by regulation.)

Everything already in your possession is accessible, and hence could be used as parts. Ask yourself this: "What might I be able to use that for?" Don't ask whether something is a "good" part. This approach is particularly valuable if you are working inside an organization and your desire to innovate comes up against boundaries imposed by existing activities or products.

What you don't already own but need is probably easy to obtain at little or no cost. The alternative is to accept that what you need is beyond your resources. This may seem counterintuitive, but the alternative principle would lead you to end everything before you even start. The Web can be to you what the photography catalogue was to Dr. Maiman. For innovation prototyping, you have access to an amazing array of resources online. With the meaning of "part" defined broadly—*whatever you can bring together at table scale to illustrate something about your problem*—all those resources can be parts for you.

Parts help you glean what might go wrong at the next scale. As you assemble parts not originally conceived to work together, you have an opportunity to become at once constructivist and skeptical. Your constructivist self finds and assembles the parts to work together. Your skeptical self doubts the whole will even work.

Innovating With What You Already Have
in a Corporate Setting

If you are innovating within a company, you may wonder whether the process is the same. In principle, everything you already have is an accessible part. In your case, this may include what your company already does, produces, and/or procures. And it includes everything routinely brought together to make your company's product(s) or service(s) real in the hands of its beneficiaries (whether consumers or other businesses).

This may help you relax the tension that may come from feeling the need to leverage your firm's core competencies while innovating—something that seems to be one of the largest hurdles to innovating in a corporate setting. It frees you up to think about them broadly: For the purpose of innovating, your core competencies are not limited to what makes your current offerings good, but include everything your company has built over years to serve those offerings effectively—as long as you can imagine how to repurpose them.

For instance, if you think of your core competency as building cars, you may be compelled to innovate on your production process or on features of the car product. But you could also think differently about everything involved in building your cars: expertise, know-how, regulations, functional departments such as finance, inventory management, procurement, and legal, and so on. Following the logic of this chapter, you may inspire and make tangible a new purpose for your innovating by thinking how these components of your firm might help you—as parts—in simulating something entirely new.

By viewing as potentially repurposed parts everything you already produce, procure, and do as an organization, you allow yourself to innovate without constraints. You may leverage what you have and still innovate inside or outside your company's current offerings.

This is how an established company can behave like a startup.

Innovating with Existing Scientific and Technological Knowledge

When it comes to innovation, individuals may feel constrained by their education in a particular discipline in the same way corporations are constrained by their core competencies. This can be especially acute for individuals with advanced degrees in a specific research field. But it doesn't have to be that way.

Education—in science and technology or, for that matter, in *any* discipline—is a powerful tool for the purpose of innovating, and is most powerful when you abandon the idea that you are constrained to a specific area of knowledge. Any constraint is actually your own doing. Education increases the depth and breadth of your thinking, equips you with new skills, opens your mind to new vantage points, and increases your literacy in a multitude of disciplines. Give your education a chance, and it will be the source of the ingenuity you need to make a problem tangible.

With that in mind, every paper or patent—even those not in your particular discipline—could be a part. So, too, could your technical training. With papers, it's a matter of seeing them for their underlying promise of offering you a means to reproduce something about nature, the world, a machine that we now know is possible—that is, a new "truth." You can use the claims and specification of patents in much the same way, but need to pay extra attention to how a patent potentially constrains your freedom to operate.

You may wonder what you might be able to do now that someone else has figured out that new "truth." You may find an entire paper useful, or only a subset of what the paper says. Perhaps what's useful is something the paper helped you think about (an adjacency). You may wonder what else might be possible, or what might result from combining what is reported in multiple papers (or patents).

You use all this to demonstrate that an aspect of *your* problem is also possible. And if your technical education taught you anything, you know it is for nature, not you, to decide whether something is actually possible. Your job is to come up with a way to pose the question to nature or come up with a contraption that might help nature give the answer you desire.

As you reason how to make the best use of a paper's contents, you'll address how to make it easy to experiment with that knowledge. You'll also likely have to figure out other material parts with which to reproduce what you need from that paper at the scale your current resources permit.

In my experience, it takes about as long to create a cursory demonstration using parts technology parts to guide your thinking in this way as it takes to produce a PowerPoint presentation. Over two weeks, working part time, the daring participants in my workshops cycle through five or six innovation prototypes—reaching out to people at the same time—and about ten variants of a problem. In most cases, getting the first demonstration going may be as easy as willingly following some online instructions.

As you progress through scale, the meaning of what's accessible will change, and so will the magnitude and impact of what you seek to demonstrate. But these principles will apply all the same. They can help you implement the basic principle with which I opened the chapter: You want to learn a vastly disproportionate, indeed *unreasonable*, amount with the money you have now, rather than postponing learning for when you have more money.

Parts scale problems down

Your first innovation prototype, like those of my examples in chapter 1, will be indistinguishable from the starting point of any "failed" company. I am asking you to resist the temptation to search for the "innovation" in the prototype, and to instead allow that search to unfold. This implies that the value of your prototypes lies in what it demonstrates is possible. This perspective is absent from most recipes that conflate *innovation* and *product*.

Your job is to layer proof of concept upon proof of concept as you scale up. What you have at every stage is but a demonstration of what you will be able to accomplish as your organization grows in the way you envision it growing. It is also a demonstration of everything you now know you will be able to accomplish as you scale up resources to the next level. So, as you embark on your innovating, you are better off thinking about what, if proven true, will best demonstrate that the next step is possible.

How do you do that? It begins with choosing what you need to make tangible and putting some parts together. You choose the aspects of the problem you need to make tangible by being honest with yourself. Of all the things that can go wrong, which would make you stop wanting to be "investor zero"? Which demonstrate best what might be accomplished at the larger scale? You choose the parts that allow you to experiment more effectively at the scale that corresponds to your current resources.

> Your job is to layer proof of concept upon proof of concept as you scale up.

The mind-set is somewhat the opposite of what you may have come to expect from the "glamour" that typically surrounds "entrepreneurship" and "innovation." The problem that gives you purpose is enticing, but it likely is poorly formulated. You *want* to prove it all wrong. You *want* to learn what makes it wrong. You *want* to give yourself an opportunity to try and err so you can determine whether there is a variant to your problem worth pursuing—one that makes it right.

Working with parts as I propose in this chapter helps you get to those answers faster and for less money. Parts are not there to help you build a product you imagine, but to help you imagine a problem solved and what a solution would need to *do* by focusing on that real-world problem at scale.

It is not about glamour. The problem that gives you purpose is enticing, but likely poorly formulated. You want to learn what makes it wrong.

Examples of Technical Parts that Represent a Larger Reality

If you've never worked with a microcontroller (or a bare-bones computer), it may sound intimidating. To me, they are an auxiliary part with which to simulate anything that needs some degree of choreography among different parts (sensors, actuators, small motors, or whatever else you hook to it).

You don't actually need to understand how a microcontroller or bare bones computer works to use one. There are myriad projects online through which to train yourself on how to use them as parts. A simple Google search will return how to add a Tesla-like screen to your car, for instance, or equip your Volkswagen van as a weather control station that can guide you to where the wind is best for surfing. Your "training" needn't take longer than a weekend. I've gotten people trained to use microcontrollers as a tool to inform their problems in about two hours.

You could certainly use a microcontroller as a component in any electronic gizmo (if that's where your hunch is leading), but there is so much more to a microcontroller. For instance, you can very easily use a microcontroller to turn light bulbs on and off and thus simulate demand in a table-scale simulation of an electrical grid. A microcontroller or a bare-bones computer might allow you to simulate an application that requires a large degree of coordination of other components. (CreepyDOL did something similar.)

Further, at the scale of a table, a large control room is indistinguishable from a microcontroller. You could use microcontrollers to simulate the effect you want from some mechanical controls. A microcontroller might even allow you to simulate tangibly some of the non-tangible aspects of your problem—distribution, time delays, and so on. Visual artists even use them to add motion to their works or enable interaction.

You can get a microcontroller (e.g., Arduino) or a bare-bones computer (e.g., a Raspberry Pi or the equivalent) delivered to your doorstep in two days for less than $50.

Another example of a multipurpose tangible part—a microfluidics kit—costs about $1,000 and can be delivered in a week. Essentially a set of small pipes, it's what you would use for all manner of lab-on-a-chip applications. And to some degree, it may be a first means to simulate what needs to happen in a larger system of pipes, or a way to use very small amounts of expensive reagents to test an idea. Conversely, you may also be able to simulate the piping with tubing you can get at any hardware store.

Although microcontrollers and microfluidics may ultimately be part of a solution, here I am inviting you to think about them as elements that allow you to emulate something else. It could be a control room, or the piping in your future manufacturing plant.

Taken further, you may also be able to simulate an entire biosafety lab in a box, and then make a business of selling field-ready biosafety labs. It may take nothing more than thinking through what needs containment and assessing whether the container, instead of being a room, could be a specially constructed box that achieves the same or even a higher standard of safety.

Placing a constraint on the scale at which you make your idea tangible forces you to pay attention to the problem. Using readily available parts is a way to scale problems down. Scaling problems down to table scale not only allows you to assemble parts more quickly, but also lets you bring problems down to human size rather than system size—where the need for statistics may make it hard to separate symptoms from root causes, if not confuse the problem altogether.

There are many benefits to thinking of parts as a means to scale a problem. It creates a concrete task. It makes it easier to see what is wrong, because at that scale what's wrong is also concrete, and because whatever can't be made tangible either is still too abstract or can be explained better as part of a scale-up sequence.

Scaling a problem with parts creates a concrete task and makes it easier to see what is wrong.

Conversely, the biggest "risk" to innovating is setting your inquiry in a way that makes being wrong unaffordable. Allowing yourself to be wrong is important because that's an important way in which you learn, but also because you are going to be wrong a lot before you find a path to being right. As a matter of fact, you need only be approximately right once: when you are getting ready to go into production.

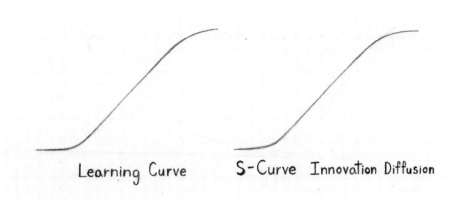

Learning Curve   S-Curve   Innovation Diffusion

We use the same kinds of curves to describe changes in knowledge and changes in adoption (as well as phase changes, etc.). In physics, whatever happens in the flat region of a curve may be causing the phase transition, but progress there cannot be measured using the same observable we use to capture the phase transition. In other words, adoption is caused by something that cannot be measured by adoption.

> Your biggest "risk" is setting on a path that makes being wrong unaffordable.

Being wrong is definitely easier to handle when the impact of your being wrong is commensurate with how much there is to lose. The mind-set of layering proofs of concept as a means to approach scale gives you that by construction: If you've focused your proofs of concept on being wrong about the things that will make your idea senseless at the next scales, being wrong is affordable and possibly beneficial. And this all begins at table scale.

In the end, all I am asking you to do is try to assemble some parts together—some may be technical and others not—in a new way for the purpose of clarifying a problem to yourself. I call that *scaling the problem down*.

Eventually, a few scale ramp-ups later, you may produce an innovation because others—unaware of the parts you've put together—adopt for use whatever it is you conceived and derives some benefit from doing so. And that justifies their calling what you did an "innovation," even if you and I know the innovation came not from the product per se but from how you brought the parts together. Remember, what made Ford's Model T an innovation was that it was "affordable," not that Ford had invented the car—the automobile already existed. Affordability had as much to do with the car as it had to do with the assembly line, the financing strategy, and raising wages—all *parts*.

Parts will tell you all you need assume. They will tell you what you may need help with, and in doing so they will supercharge your imagination. Once put together, the parts illustrate the problem, the scale, or the verification. That prepares you to discover where the opportunity for impact lies and how to ramp up through scale.

And again, all along the way, nothing prevents you from using mostly parts you already have.

Takeaways

• You can get started right away. Assemble some parts you either already have or can easily get. Acquire knowledge as you go. Use parts to make your problem tangible. A part is good if it allows you to demonstrate a function or an aspect of your problem.

• *Anything* to which you already have access may be a part. If you are innovating in a corporate setting, this includes everything you currently do, assemble, or procure. For everything else, there's the Web, where you can find all sorts of inexpensive parts with which to emulate even the more non-tangible aspects of your prototype: from the technical to all sorts of services and products you may use to address legal, organizational, and manufacturing aspects of your problem. There is also a wealth of available resources online for acquiring knowledge you'll need.

• As you bring parts together, questions such as "What else might I use it for or with?" and "What if these parts worked together?" and "What problem would they illustrate?" help you more than asking whether something is a part and whether it's "good."

• Choose parts so you can reason about your problem with your mind and your hands. Choose parts to make *any* aspect of your problem tangible—even a solution. Think about parts as elements in a conversation you are having with your problem. Parts respond when you ask:

  What makes this hunch so special?

  What could go wrong? What do I have to do to show that it is wrong?

  What do I have already that I can use to demonstrate what my hunch may be missing?

• As you assemble parts, they'll tell you what you need to assume at the current scale, what you aren't seeing, what you are missing, and whether the next scale is even possible.

Parts that demonstrate a larger reality

Parts to enact and emulate non-tangible
aspects of your problem

A working Innovation Prototype
organized to demonstrate a problem

Parts that help scale down the problem

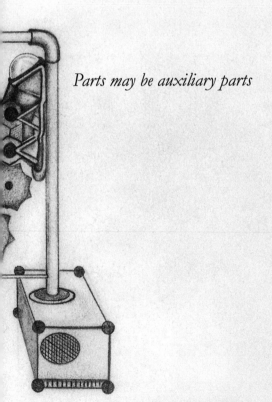

Parts may be auxiliary parts

*enacted by parts and people and
at the scale of the resources you have*

INTERFACING WITH PEOPLE

You will need to interface with people as you innovate. People will influence the direction your innovating takes. Who those people ought to be and what roles they may play cannot possibly be known until they begin to influence the problem you're solving. This has elements of a Catch-22.

There are two common ways around this catch. One is to decide on the people and solve for the innovation—as designers often do. Another is to fix in advance what your innovation will be, get a team, and solve for the market (which is people)—which is what marketers would have you do. I propose a third: Focus on the problem and solve for people's contributions in the form of information, capabilities, and skills.

In the third approach, people are as much a variable in your innovating as are parts. Nothing is fixed. Your eventual innovation is as much an assembly of parts as it is an assembly of people, and your deliberate interactions with other people will shape your innovation as much as putting parts together will. In fact, just about everything about parts in chapter 3 applies to people—even though people are not parts. For instance, as you innovate, some parts stay, some parts go, some parts inform the next step, and all parts inform the problem, helping you to make it more and more tangible. Likewise with people: As you innovate, people will come and go in the same way, providing information you need or skills you may be lacking. Some people will inform the next step of your innovating.

All the people with whom you interact will inform the problem, helping you make it more and more tangible.

Parts help you understand how your problem functions, and inform feasibility and how to scale up. But there is absolutely nothing in parts you put together in a gizmo that can inform what only people can inform: *impact*. Specifically, people can help you understand how and why to bring parts together for impact, and the scale you need to attain to make that impact tangible.

Sooner or later, it will become clear to you that you need people. There is a tendency to jump way ahead and think you need to design for users or acquire some board members, or a business partner, or a technical guy—even when you haven't yet created an organization. That's backward thinking—often confused with "working backwards" from the objective. Better to focus on getting the information and skills you need at a given moment, without worrying about whether someone is going to stick around to serve on some kind of board that may not even be relevant today. To be sure, there may well be a team, partners, users, advisors, buyers, distributors, markets, and board members along your future path. But setting such roles in stone is a far cry from pulling people together for information and skills at this earlier stage. Besides, teams rarely come pre-assembled, which is why the notion "Get an idea and get a team" mires you in hindsight thinking.

Talk First, Count Later

We fall in love with an idea. We love it as a machine. We love it for its æsthetics. We persuade ourselves that every challenge to adoption can be assumed away, rationalized into oblivion, or rejected outright because "they" *just don't get it*. Soon enough, there is no one left who cares or even understands why you are so wrapped up with your little pet project. This is the same kind of problem as having a basement without an egress. You ought to be able to do something about it.

How? Let me begin with an admittedly ridiculous example: Suppose you want to know whether a beverage that is across the room is cold. You would most likely not sit there reasoning your way into a temperature figure. Brainstorming with colleagues will get you only so far. Chances are you will not bury yourself in arcane temperature research reports. You may even dissuade yourself from Googling your way to an answer for the average beverage someone else claims to have measured on a whim, or some other statistic. Instead, you will probably walk across the room and measure the beverage's temperature yourself. Whether you use a thermometer, use your finger, or take a sip is a matter of style, the precision you require, the purpose of the measurement, and your tolerance for risk.

Something similar is true of conversations. The information you need resides in people, not in a head count or some other statistic. The finest statistics are things like up-to-the-minute sales reports or final election results. By the time such statistics are available to you, you will have advanced so far in your endeavor that you will dismiss all other input.

If ultimate statistics are not an option available to you, then you have a trade-off to explore: You may seek proxy statistics and find comfort in numbers, or you can walk across the room and just start talking to people about what they're doing. What you really need to understand is the structure of the space you are thinking of getting into—and that understanding comes from people. You could still conduct a survey to compile statistics from those people, but that's one of the most inefficient ways of reaching the kind of understanding you need considering all the uncertainties at this point.

You can get closer to that understanding through a conversation—with an industry expert, a potential user, or just about anyone who might have something useful to say. It turns out that a conversation is a way out of this particular "basement without an egress." Later, you can decorate what you've learned with statistics, narrow as they may be, that shine light on the critical aspects of your idea.

In other words: Talk first, count later.

> People come and go as you innovate, providing information or skills you may lack. Your eventual innovation is an assembly of people.

As if one Catch-22 were not enough, it turns out there is another caveat: The people who will ultimately adopt the outcomes of your innovating do not yet exist. Sure, those individuals may be living, but they aren't yet the people they will be when the time comes for them to derive a benefit from your innovating.

It takes a non-insignificant amount of time, often years, for an innovation to emerge and to be adopted. A lot can happen in that time. The people you interact with may mature new ideas, and you have a role to play in that. Consider smartphones, which were but a niche product in the mid-1990s and a far cry from the mini-computers and lucrative platform for the app business they are today. If you had told your grandfather back in the early 1990s to hold on because you wanted to take a picture with your phone, period handset in hand, you would have been dismissed as crazy for merely *suggesting* such an idea. "There's something wrong with that kid," your grandfather might have said.

Now consider the problem that gives you purpose. From the previous example, it follows that—today—you are wrong.

As you scale up, you need a way to interface meaningfully with people despite knowing that they cannot possibly deliver the straightforward answers that would turn your innovating into a mere exercise in template matching or a wholly user-centric design. There are several things you need to learn from interfacing with people, not the least of which is the organizational model that will facilitate that scale-up—that is, the way the people who join you will work together within an organization that has yet to emerge. Please note that merely founding a company, a non-profit, or some other kind of entity does not an organization make. A company is a vessel that, absent some organizational principles, does not move. The converse is also true: An organization alone is nothing more than a bundle of organizing principles.

As they mature, innovation prototypes *accrue* people incrementally. The degree to which these people continue to share an interest in the project determines their roles, how long they play them, and the depth of their interactions with the project. These interactions will influence and possibly change the problem you are solving.

So, how do you work with this variable, *people*? How do you deliberately or serendipitously attract them to your innovating, use what they have to offer, and move them in and out of appropriate roles, understanding that as with parts, people will come and go?

Essentially, these people will fall into one of the following categories.

People may have *information* you need. This information may be about something in your prototype, such as one of the uncertainties or unknowns you had set aside for later (as discussed in chapter 2). It may be about parts— say, the parts that simulate distribution. Or it may be about other people to whom you need to reach out, such as experts or potential users.

People may have *skills* you need, generally dictated by the parts of your prototype. For instance, you may lack the skills to work with the environmental regulations (a part) for your distributed chemical manufacturing idea, and so some help from an attorney and perhaps the right kind of engineer may be in order. People from whom you acquire information may also help identify skills you might need.

People who have information and/or skills may, at a later point, become contractors, consultants, advisors, core team members, and play other roles in the organization that is emerging as you scale up your

> Serendipitous encounters may yield information, skills, or money you need.

prototype. For now, though, all you can really assess is their knowledge, capabilities, and skills. Any longer-term relationship will hinge on that.

People may have *money* you need, too. At a later point, they may become users, investors, partners, buyers, and so on. As I discuss in more detail in chapter 8, we tend to look for the money first, or at least far earlier than is necessary, which is very disruptive of innovating. In fact, time spent searching for money could almost certainly be spent more productively doing something else—and for that too, there is a way to view your fundraising as an incremental step for which you've been naturally preparing through your interactions with people.

The way these people affect and perhaps even fundamentally change your innovating determines the roles they may play in the future. You'll seek them out for specific roles based on how they've affected your innovation prototype. As an operating principle, you want to work with people who routinely broaden your understanding of what you're doing and help you reach beyond where you're at to get more people and information—no matter their possible future role.

As you scale up, and more and more people are involved in a circle of ever-changing relationships, you may reach a point where some of the people gel. That's when future roles become clearer: core team members, advisory board members, financial advisors, whatever. You select people because they have what you need to push things forward. And you understand what's needed to push things forward because you've understood that from your interactions. They—the people and their interactions—become an organization, which early on you may call *the team*.

This effectively allows you to view both organization and team as emerging from your prototyping and scale-up. And, as with parts—that is, the non-people tangible aspects of your innovating—non-linearities play to your advantage: At a sufficiently small scale, every one new person (an incremental change) may fundamentally change the destination of your innovating (a disruptive change).

A team, thus, is a construct that evolves as you engage in the activity of reaching out to people for a variety of things. And there is something interesting to note about the emergence of a team compared with going out to look for money: As you build this circle of people that strengthens your innovation prototype, you are in effect assembling a community around your future innovation. Indeed, money may *come to you* from people who want to "buy in" to your innovating. They've all been exposed to your innovation prototype—directly or by referral—through your quest for information or skills or, more generally, advice.

Some may want to "buy" a portion of your scale-up and hence a role as an investors. Others may want to buy a version of what you're actually creating; these may be the "lead users" spoken of in business parlance. Still others may buy a portion of your innovation prototype and stay on as partners.

There is yet another category of money that does not come to you so much as you go to it. Investments from venture capitalists, angel investors, private equity folks, and others may come to you as a result of your quest for information, skills, and advice. In chapter 8, I discuss how you can come to view it as a natural extension of your inquiry.

Finding people

How do you find people? Consider information. You are, like everyone else, already a part of many communities. One or more of those communities has likely given you information already that led you to develop the problem you think you can solve—that is, that led to your hunch. From wherever you start, reaching people means reaching beyond the community or communities in which you developed the hunch. You are not a community monogamist. Your life is a bunch of ever-expanding circles and communities. The obvious way to begin to reach beyond is by communicating with the people you already know. You may choose people to communicate with based on the kind of information you believe you need, or whether you trust that they will tell you outright why they think you are wrong (if they do). What you need is going to change as a result of having those conversations.

That covers at least the set of people you can *imagine* have information you now know you need. You can do all this through your social media connections, targeting specific interest groups that are particularly aligned with your purpose, or through "cold calls" to thought leaders and other targeted individuals—to name just a few.

However, at a sufficiently small scale, the merits of being highly strategic about your inquiries are at best dubious. You have a hunch. You have more questions than answers. Most likely, one question is more pressing than others, such as: "Do you know who knows something about ... ?" That's a question that might be answered by people you know.

That leads me to the other set of people: the ones you meet serendipitously and who volunteer information about what they do that ends up being valuable to you. This is all about chance encounters. It turns out that the people who move in your same circles have more going on than the limited "personas" you may attach to them. They have more dimensions, and they are part of multiple other communities of which you may not be aware. So, in addition to all the "profiling" you may do to figure out who you know already that can help you, you can also make it your business to engage actively (and in person) with people without any other agenda than to allow yourself to be surprised.

I am encouraging you to participate in social groups, to attend events, and to open yourself up to casual conversations while the after-event cocktails are served, even when the event does not seem to be directly related to your innovating. Without asking, there is no way to tell whether the person standing next to you at the member event at the museum knows something you don't but should. You just need to be ready to listen and have a casual description ready when your turn comes to answer the "What are you up to?" question.

> Chance encounters are valuable. Until proven otherwise, everyone you speak with is a potential expert.

Chance Encounters

When it comes to finding people, you can go either way: you either assume you already know the kinds of people you need to meet and talk to, or you take for granted that you do not really know what you are doing and assume that every encounter with other people stands to make you smarter. Opening yourself up to chance encounters may seem obvious, but doing so has become anathema in how entrepreneurship is typically taught. Instead, the focus is on creating the ideal list of people you should talk to. That takes up so much time that chance encounters—and their value—become an afterthought at best.

Driven as you are to arrive at the destination you envision, it is incredibly easy to limit yourself to interactions with people and in communities that are directly related to your mission in some obvious way. Indeed, you meet people by participating in events that interest you and them. However, no one (with the possible exception, by omission, of textbooks on marketing personas and the aforementioned entrepreneurship courses) ever said that you can't find a person relevant to your inquiry at an event that has nothing to do with your mission. At an opera performance you may come across someone who knows nothing about your problem but knows someone who does. Much the same could happen at a museum opening, at a dinner at a friend's house, at an event at your child's school, or at an airport where you find yourself stranded by inclement weather. In fact, you can meet people relevant to your inquiry *anywhere*.

At the risk of stating the obvious, all it takes is having a conversation that begins with a casual "Hello," asking people what they do, and having genuine interest in the answer. When your turn comes to answer the same question, you need only offer a casual description of what you do and what you are after. Humans are reasonably good at conversing—better than they are at answering questionnaires.

Indeed, limiting yourself to just the *obvious* interest group and the *personas* you profile sounds as boring and narrow-minded as confining yourself to the basement portrayed in so many entrepreneurship books.

For some extra motivation, here are some notable examples of significant casual encounters:

• Sergey Brin and Larry Page, the founders of Google, met when Page visited Stanford University as a prospective student.

- Steve Jobs and Steve Wozniak, the co-founders of Apple, were introduced to each other by a neighbor.*

- Bill Hewlett and David Packard, whose giant technology company originated in a Palo Alto garage, reportedly met when they both joined Stanford's football team as freshmen.**

- It was at a family dinner that a guest speaker in my class figured out that he ought to introduce his new technology to an industry he had initially dismissed. A long-time family friend who worked in that industry happened to be at the table and told him he was wrong. The company was sold two years later for nearly $100 million.

The lesson here is that the person who can give you good advice and perhaps even join you in your adventure is someone you've already met or someone with whom you may have a chance encounter at an event at best tangentially related to the problem that gives you purpose. The serendipitous nature of this comes as a relief. After all, with nothing but a highly volatile hunch to go on, finding the perfect event to attend might be next to impossible.

*See Jason Hiner, "Apple's first employee: The remarkable odyssey of Bill Fernandez," *TechRepublic*, September 2015.

**Ed Sharpe, "Hewlett-Packard, The Early Years," www.smecc.org/hewlett-packard,_the_early_years.htm.

In your ambition to scale up, you may feel some urgency to make generalizations about the people you meet as you listen to what they say. But resist the temptation to categorize them as some archetype you think you need. Instead, see them as the individuals they are and open yourself to learning from the unique information they may provide.

Put another way, these are genuine people you are talking to, not instances of a distribution. Your purpose is not to figure out now what scale looks like in terms of number of people, but to figure out something from individuals' experiences and the specific information they may provide that will improve your hunch, refine your problem, and help you reconceive what you prototype for impact.

You need to gauge impact, and you can only do that with people. Your prototype needs to speak to something that actual people will do for their own benefit. It ought to be possible to relate what you propose doing to how other people have gone about solving problems you find to be similar. To understand impact, you need to engage with people. In my experience, most individuals respond best to conversations and to decisions, and are notoriously bad at judging ideas on the spot, brainstorming about hypotheticals, or identifying what they may need several years from now. To be *productively* wrong you need the input of others, and your interests are not well served if both you and your counterpart are guessing.

Fear and Data

When it comes to factoring impact into your innovating, we often fear getting tangible too quickly. In truth, though, the "fear" is lack of knowledge—just as with parts. The response to that "fear" is often to rush to Google seeking immediate relief in some figures quickly pulled together.

Few of us realize that those figures are as theoretical a prototype of impact as is any imaginary product.

If looking for relief in numbers is the entire process, you will arrive at an existing market and the "innovation" is likely to be a copy. There is nothing inherently bad about a copy, but producing a copy may not be what the innovators originally set out to do.

Lack of knowledge creates uncertainty, and numbers alone do not reduce uncertainty. Numbers are nothing more than what we use to communicate some kinds of easily compressible data. Those data and the inferences you draw from them can reduce uncertainty. I am not proposing that you shy away from numbers; just remember that the tangible story around those numbers is what truly matters.

In the case of impact, what is tangible stems from what you learn from conversations with people, not from how many conversations you have.

All you need to do is prepare your innovation prototype to inform a casual conversation.

Conversing with people

As you go about innovating, there is no need for you to adopt some affected mode of conversation. In encounters with other people, chance or planned, it's quite common that sooner or later an opportunity will present itself to discuss what each of you is working on. That's an invitation to learn. Whether you're having an informal conversation with a stranger or with a noted opinion leader, you should feel free to ask directly about what your conversation partner does, about a piece of research, or something about his or her job. Until proven otherwise, everyone you speak with is a potential expert.

You learn through the interaction, not through matching the individual to some kind of template or surveying them with a standard questionnaire. And if during a conversation you find yourself wondering "Is now the time for the questionnaire?" the answer is No. In fact it is *never* time for the standard questionnaire. Questionnaires and surveys do serve a purpose, but they yield nothing like the information you may obtain through conversation. (I touch on the topic of questionnaires in chapter 12.)

As you converse, ask questions that go with the flow of the conversation and that help give you a detailed understanding of what your counterpart is talking about: "What made you think of doing that?" "When did you start?" "How many people did it take?" "Who did you have to find to help?" "Where did you find them?" "How did you go about [this or that]?" "What kinds of

Prepare your innovation prototype to inform a
casual conversation. You don't have to share your
complete idea.

difficulties did you encounter?" "Why do you think it worked, or didn't work?" "How did you pay for it?"

Be mindful that questions such as "How much would you pay for a such and such with these bells and so whistles?"—that is, questions aimed at garnering quantifiable evidence—usually draw rather uninformative answers. If you really need a figure, I advise that you follow the teachings of the *Hitchhiker's Guide to the Galaxy* and use 42 ("the Answer to the Ultimate Question of Life, the Universe, and Everything"[1]) and then focus your efforts on finding out what the units ought to be until you know what you really need to know.

Leave figuring out how the experience of the person you talked to is relevant to your own innovating for later. First you need to understand his or her experience. Resist every temptation to seek approval for your idea. That is not the conversation's purpose. If you detect some parallelism with your idea, turn it into a question about your counterpart's experience, not about your problem.

When your turn comes to explain what you are up to, use what you have learned. You learned it as you worked to make your own inquiry tangible. Discuss specifics. Remember, though, that this is a conversation, not an opportunity to spill your guts with a half-baked "elevator pitch."

You have a problem you envision solving, you have a cursory idea of impact, and there is something tangible you are specifically doing. Talk about all that. There likely is something very specific you need to know at this point—your uncertainties and unknowns—that you can bring up in the conversation, because your counterpart may know someone who can help. A friend of mine once told me that in the early 1990s, when she was starting her company, she asked all the people she met whether they knew anyone who worked as what we today call a software engineer.

At some point in your conversation, you may begin to worry that you're sharing too much about your idea. That issue always comes up in classroom discussions. It's true that what we don't share can't generally be copied, but it's also true that what we don't share cannot elicit a response. So, there is a trade-off to explore.

You don't have to share your *complete* idea. There are ways to present what you are working on with details removed or through an example. It is to your advantage if these conversations are easier to have, and that can be helped along by an analogy or a "near miss."

Near Misses

I am using the term "near miss" in a constructivist way, rather than in the catastrophe-averting sense in which you may be accustomed to hearing it. The inspiration for this use comes from the work on artificial intelligence of MIT's Patrick Henry Winston.[*]

A near miss can be the conversational analogue of using a strawberry in lieu of a bacterium. For instance, you may be interested in using bitcoins to set up a payment-clearance service. That might be hard to explain to your conversational counterparts but easy for them to copy. A near miss in your conversation might be to use Square or Apple Pay or Google Wallet as an example.

[*]See, for example, Patrick Henry Winston, "Learning Structural Descriptions from Examples," Ph.D. thesis, Massachusetts Institute of Technology, 1970. A shortened version is in Patrick Henry Winston (ed.), *The Psychology of Computer Vision*, McGraw-Hill Book Company, 1975.

You stand to get better responses if, instead of explaining how everything will work, you focus on what you think you want to accomplish.

Being wrong; prototyping impact

You can only hope the people you converse with will tell you that you're wrong, because a veritable treasure trove of information may lie behind such a statement. Near misses—in which you are "wrong" by default—will probably help get there.

Just because you are told you are wrong, though, you shouldn't necessarily believe everything others say.

Once you have all this information from other people, there is the question of how you incorporate it into your thinking. For that, you have parts.

People can tell you facts. They can point to information they're surprised you did not mention when outlining your idea. They can supply you with these tidbits from their own experience that seem to contradict aspects of what you shared. They'll likely state these as evidence that you're wrong. All of this is so useful that you ought to practice how to answer the "What are you up to?" question so your conversations unfold this way.

But people can't really tell you what makes your entire idea wrong. Finding *that* out requires you to bring what you learned back to the realm of parts and "debug" your innovation prototype. You do that by specifying anew what your parts do, moving them around, adding new parts, and/or removing some parts altogether.

Ultimately, you are bringing parts together to right the wrongs you've been told about. Your purpose is to find out what made your explanation appear wrong to your conversational counterpart. It may have been the presentation itself, or your problem, or something in between (parts, people, scale, whatever).

You want other people to tell you that you're wrong, and you want to probe that judgment through conversation.

Operating Principles for Conversations

· Everyone you speak with is a potential expert. If you persuade yourself in advance that there's no information to be gained, you will not gain it.

· If your idea was so easy to explain to someone that a casual conversation was all a person would need to run with it, it invites the question: Why didn't you just run with it?

· Focus on the problem you are trying to solve and how you are going about it. Set up the conversation to ask or answer questions.

· Your explanation to your conversational counterparts should be just engaging enough to make them want to know more.

· Ask questions directly; don't preface them with how you got to those questions. Save that for later, when you're asked for more information. My data show conclusively that most people want to hear only three new things at a time.

· There is always a way to ask a question that requires only minimal exposure to your idea.

· You are not seeking approval or a pat on the back. A focus on why your idea is *good* will likely result in your getting little or no information.

· Giving specifics with certitude makes it easier for people to find reasons in their own experiences for why they think you are wrong.

· Your conversational counterparts will likely reveal information, and eventually share skills, expertise, or capabilities, only if they feel the conversation was genuine and substantive.

· If you are offered advice, be sure to capture your counterpart's rationale.

· You can follow advice only when it makes sense to you. That may not happen immediately. The person who gave the advice to you saw something relevant before you did, and may have more to offer.

The people with whom you speak have an intuition—a "verification recipe" borne of their experience—that they use reflexively to judge that something cannot solve the problem they imagine you want to solve. In the end, your own verification recipe to judge that the problem *is* solved will be informed by elements of these other recipes. This is exactly why you want other people to tell you that you are wrong, and why you want to probe that judgment through conversation.

Takeaways

- You have an idea. You want to be told you're wrong. But you don't want to tell *yourself* you're wrong, because that will become a self-fulfilling prophecy and you won't do anything more. Therefore, you need other people to tell you that you're wrong.

- To help other people tell you that you're wrong, you need three things. First, you need a conversation starter—your response to the question "What are you up to?" Second, you have some specific things you need to know—your uncertainties and unknowns. That knowledge could come from your conversational counterparts. Third, you must have a genuine desire to listen to what others are up to.

- You need to resist the urge to categorize people you talk to for the purpose of counting them. Your purpose is not to figure out now what scale looks like in terms of people.

- You're talking to humans, and humans make terrible robots. Humans are incredibly bad at following the instruction set of your preconceived archetypes. But you can figure out something from their individual experiences and from the specific information they may provide that will improve your hunch, refine your problem, and help you reconceive what you prototype for impact.

- Your innovation prototype will accrue people through these conversations as users, team members, and so on.

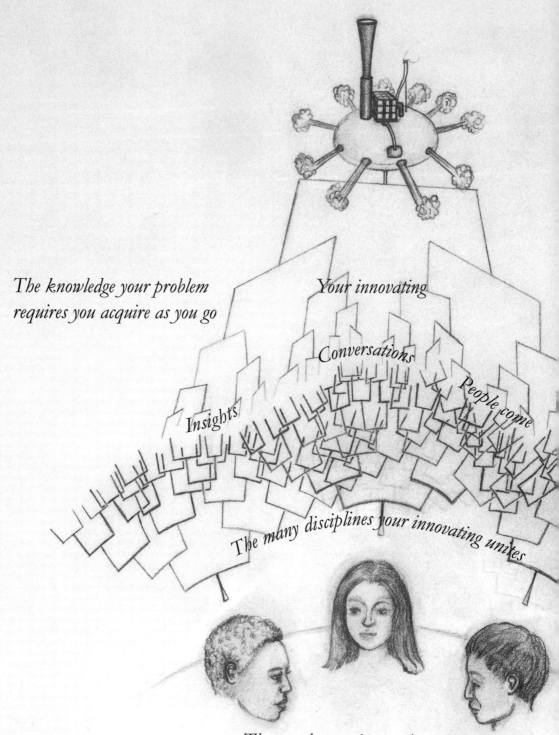

The knowledge your problem requires you acquire as you go

Your innovating

Conversations

People come

Insights

The many disciplines your innovating unites

The people your innovating encounters

– INTERFACING WITH PEOPLE

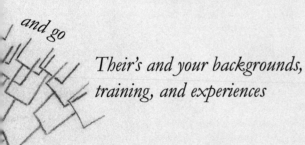

and go

Their's and your backgrounds,
training, and experiences

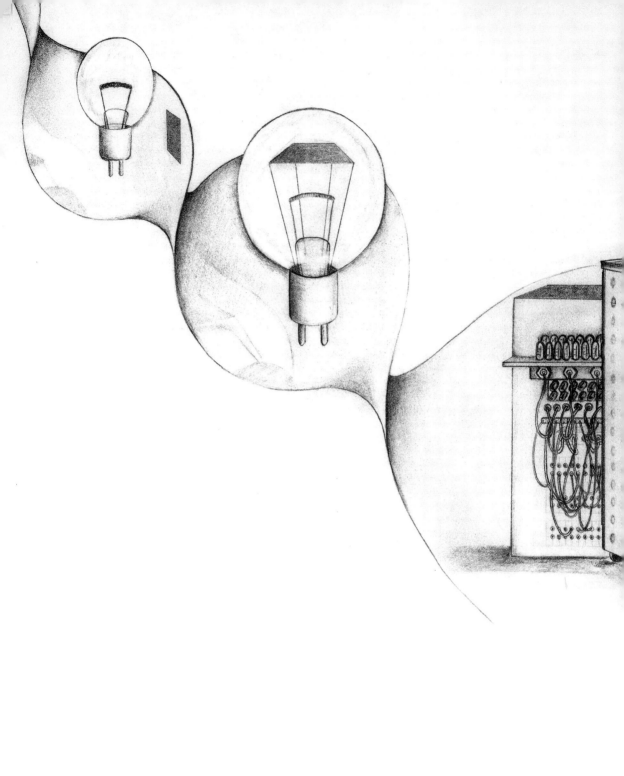

5

AT A SMALL SCALE, NONLINEARITY IS YOUR ALLY

The smaller the scale, the bigger the space you can explore.

You can turn nonlinearity—one of the culprits responsible for the frightening things about innovating—into your advantage. This is what chapters 3 and 4 prepare you for: essentially, to explore an increasingly diverse set of larger realities by systematically adding, discarding, and recombining easily accessible parts and insights or skills brought to you by people. That's how you learn far, far more than the resources you have would suggest is possible.

In this chapter, I tell you how to use nonlinearity to your advantage. Specifically, I tell you what this nonlinearity means to you, how to exploit it to make your idea more robust, how it helps you explore impact and scale, and ultimately how you can develop your own sense of progress absent the convenience of a linear indicator of progress. No such indicator actually exists.

What is nonlinearity?

As I note in chapter 1, the process of innovation is nonlinear. That means you cannot count on things being easy to forecast. I use "nonlinear" to mean that small changes at the outset of your process may have unpredictably large effects on the outcome. The converse is also true: Slight changes in the outcome you envision may imply dramatically different starting points.

Nonlinearity also implies that enacting the process of innovation requires that you and your evolving hunch undergo some form of recurrence or iteration through which the problem you solve becomes increasingly clear and nuanced.[1] Some call this approximation, relaxation, multivariate optimization, chaos, and other things, but what's important is that unless you have pre-specified a solution there likely are many solutions, and getting to them will require that you know when to stop.

The smaller the scale, the bigger the space you can explore.

If you are a physicist or an engineer, you likely are already familiar with the idea that "rules" can be different at different scales. You see this in quantum versus classical mechanics, or microeconomics versus macroeconomics. Chemical engineering as a discipline exists precisely to deal with differences at lab scale versus factory scale.

In innovation, too, "rules" can be different at different scales. Seemingly small changes to the problem that gives you purpose may translate into dramatic changes in impact. Similarly, seemingly innocent changes to how you envision impact may fundamentally change the nature of the problem and possibly help you overcome challenges that appeared insurmountable.

Pilots

I am often asked whether people should run pilots of their ideas.

I think a pilot is always a good idea; however, pilots—like products—are already very sophisticated constructs. Running a pilot is generally a significant project. It requires lots of resources and a conscious commitment to an idea. But even if the time is not right for a pilot, the pilot *mentality* is the right mentality to have from the very beginning, when all decisions are yet to be made.

How do you achieve that mentality? You bring the problem down to a scale that's likely smaller than what a pilot would entail. At that scale, which I've been calling table scale, the purpose isn't yet to assess the scale-up needs. Rather, it is a way to discover a direction and a path to scale-up that you can think of as a sort of sequence of pilots.

It helps to think of pilots as intermediate checkpoints that help you correct course by "linearizing" the challenges and the risks that lie ahead.

You may already be experiencing this nonlinearity without realizing it. After all, that innovations can be disruptive and that disrupting technologies/innovations/products are at first underperforming the incumbents are both outcomes of nonlinearity. If you take those facts as givens for innovations, this is what you may be implicitly accepting about their origins: disruption is an outcome; you can only speculate *what* will be disrupted,

In innovation, as in physics, "rules" are different at different scales.

whether it will be disrupted, and *how*; and that at first your idea will underperform everything else out there. It follows that what you start with is incremental—that is, not disruptive—and underachieving. And so, surprisingly, what you set yourself to build may not need to be better now than what already exists; it simply has to be worse today than what it will become tomorrow.

Nonlinearity need not be frightening

The fear of nonlinearity might push you toward wanting to specify as much of what you will do as you can and as early as you can—in other words, to paint a rosy deterministic future. There is no guarantee that doing so helps in any way, but it's still what most people do.

Predicting a rosy future is unnecessarily limiting; you need only ensure that at least *one* future exists. You can explore a lot if you let *future* be a variable and you constantly revisit your variables and your assumptions. And if you strive to demonstrate tangibly to yourself first that you will not fail in a predictable way when the time comes to get to the next scale, you will ensure that at least one future exists.

The way I see it, this nonlinearity is the best news ever:

- The combination of mostly existing parts can lead to something that's game changing.

- At the right scale, changing small parts allows you to explore an incredibly broad space of opportunities for impact.

- Everything you learn through those changes prepares you to develop a robust organization that operates on an entire space of opportunity, rather than focusing on a single product or technology.

- Operating on an entire space of opportunity increases the chances that at least one future exists.

I cannot imagine how you would ever be able to explore as much as you need to were it not for this nonlinearity.

In 1879, Thomas Edison completed the first successful test of a commercially viable electric light bulb. His light bulb was essentially a filament sealed inside a glass bulb. The glass bulb provided a vacuum. Edison filed for a patent on November 4.

From 1875 to 1883, Edison reportedly experimented by inserting all sorts of things into the same vacuum. One thing he discovered was that, if he placed a cold electric plate inside the vacuum, electrons would travel from the filament through the vacuum and would create a current in the plate; however, electrons would not travel in the opposite direction. Reportedly, Edison saw no value in that, but he still went ahead and patented it. Sometime around 1906, Sir John Fleming applied what Edison had discovered about the current and developed the first vacuum tube, which he used to demodulate radio signals.

The first vacuum tube allowed current to flow in only one direction. Today, we call that a diode. Early commercial "Fleming valves" appeared in the late 1910s. Until the advent of the transistor in the 1950s, the vacuum tube was the basic building block for all electronic devices. It is still used today in high-end audio applications and other contexts.

It all began with a light bulb and a thin metal plate—two readily accessible parts at the time. Although the change was small, its impact was, arguably, disruptive. It launched the field of electronics.

Google provides another example. In this case, the power of exploration came from a change in parts following insights that emerged from people. Many attribute Google's success to the simple realization that the same algorithm so powerful for ranking websites could also be used to rank advertisements—a small change. Reportedly, that small change yielded a dramatic innovation in the business model that solved a problem that had grown to epic proportions: The Web had become a nightmare full of poorly performing banner ads. Whether that's how the Google story actually happened or how it is remembered in hindsight is less important than what the story implies for your innovating. You can easily imagine a casual conversation with another person leading to these seemingly unassuming questions: "What if the search is not for websites but for ads? Could you guys search for that?"

The strength lies not in the questions but in recognizing that it is such a small change that entertaining and eventually verifying the possibility is easier than any attempt at rationalization.

These two examples point to an important aspect of nonlinearity: Replacing a part or taking to heart an idea that emerges from a conversation isn't enough in isolation. To derive a benefit from this nonlinearity, you've got to be able to see your innovation prototype both for the small-scale assembly of parts and people it is and for the larger reality it illustrates. As the Edison example shows, without that "duality" it may otherwise take years and someone else to answer the question of what a new combination (of parts and people) demonstrates is possible.

Nonlinearity's effect on mind-set

It may take some getting used before you can fully appreciate the power that comes from leveraging the nonlinearities of innovation to explore an entire space of opportunity. As simple as the process of exploration is—that is, replace readily accessible parts with other readily accessible parts and embrace or discard insights emerging after a conversation—there is nothing obvious about why it works until you've practiced it a few times. But imagine what becomes possible: You set yourself up to discover what makes you wrong so you can keep on changing seemingly innocuous parts and become ever more aware and knowledgeable about the problem you are solving.

Nonlinearity makes exploring inexpensive. You needn't gamble it all on one future but can explore an entire space of opportunity.

The idea that this kind of inexpensive exploration can yield powerful insights with which you can conceive and plan many large-scale systems is somewhat paradoxical, but it is indeed one of the most powerful instruments in your inquiry. It helps you to implement the principle I introduce in chapter 3 to learn a vastly disproportionate and indeed *unreasonable* amount with the resources you have, rather than postponing learning for when you have more resources, and it reduces your need to gamble on one specific future.

With this mind-set, what's important about parts is that you can swap them out for new ones that cast a better light on the problem and that leads to new people; everything else about parts is accessory. Similarly, what's important about conversations with people at the current scale is how each conversation stands to change the way you look at a problem and points to new parts; you decide which insights derived through conversation to keep and which to leave for later. You are not designing for specific people any more than you are imposing specific parts on your problem.

Disciplinary blinders

My terminology throughout the book is crafted to help you overcome one of the difficulties associated with innovating. Like me, you may have been trained in a specific discipline. It is hard to let go of that investment. Yet by the time you are done and successful, your innovating will have emerged from an amalgam of disciplines that no one had put together in quite the same way before. How did that happen? You must have stepped out of your discipline at some point. You might as well resolve to do that from the outset.

Your innovating may require you to bring together parts in the way Edison did, undeterred by whether the conventional wisdom advises you that those parts do not belong together; it is reported that Edison may also have tried with a feather. Or it may require that you be alert for comments that seem off the mark at first, as the idea of searching for ads did. Some great ideas may sound preposterous at first.

Trying to combine seemingly distinct parts and asking yourself "What might be possible *now*?" may appear to be somewhat unorthodox. There are, however, few other ways to allow yourself to bring "preposterous" combinations together. Our education in specific disciplines—the praise given in school to multidisciplinary thinking notwithstanding—is what makes the proposition sound absurd. Yet it is no more absurd in spirit than the belief that led the founders of Greenpeace to charter a fishing boat to stop the United States from testing nuclear bombs. There is no guarantee that your preposterous idea will ever be recognized as anything other than preposterous; rather, it is largely a matter of how driven you are to find a successful path to scale and how malleable you allow your problem to be. And that hinges upon your not mistaking the power of your imagination for proof.

Impact

There is a reason why this kind of exploration works: In the end, the only thing that matters is impact.

Impact does not stem from the parts you have, how new they are, what people say, or how many of them fit your preferred archetype. It stems from whether the community that experiences the problem you are solving is able to accomplish new things after you've solved it. Surprisingly, the nature of the problem doesn't seem to matter; rather, what matters is what happens after the problem is solved. Perhaps that's why, as I discuss in chapter 2, no one ever defines problems directly.

You can rationalize whatever happens after the problem is solved as value, empowerment, value proposition, benefits, or satisfaction. Economists use the concept of progress. When the solution involves technology—here I'm using the definition of technology that launched MIT: something that "serves to extend the dominion of mankind over nature"[2]—they call it technological progress, which people liken to technological innovation.[3]

After the fact, the *magnitude* of the impact may be gauged by number of people, or by sales, or by increase in gross product per capita, or by health, or by productivity, or by number of citations, or by the reputation of an award. But *actual* impact stems from the fact that those people—however many there are—can no longer fathom a world in which the problem you solved was a problem.

It is easy to get sidetracked by numbers. It is even easier to aim for *actual* impact: You need to think about everything you've been learning from your conversations with parts and people as a whole, and imagine how the lives of others might improve as a result of the outcomes of your innovating. Because imagining the lives of others may be difficult, you can start looking for inspiration in other stories you know about how outcomes have positively affected people's lives (to use them as parts if you will), or even begin with your own life and the lives of members of your team. What "magical" things might you really be able to accomplish after you are done with your innovating?

Only *impact* matters: people can't fathom a world
with the problem you erased.

Progress as learning

The fact that the process is nonlinear does not imply that you have to dispense with observing progress. It simply means that you should not expect to measure progress as a function of what you are building. Rather, it is your understanding of the problem that progresses.

You can observe learning as a function of the precision of the questions you pose to parts and people, and in the ease with which you recombine

:s or insights. You can also observe learning in any of the measures of progress that are generally associated with the kind of learning you do by practicing.

When you've progressed so much that it seems that you get things right more often than you get them wrong, or when you feel that there is a set of ideas and combinations that stubbornly refuse to prove themselves wrong, you can take that as an indication that the scale you are at has become too small.

The effects of nonlinearity on time

Time and resources are, arguably, the two most precious commodities as you evolve your innovating toward impact. Both are affected by nonlinearity. Mastering both can be greatly helped by managing the scale at which you operate. I have discussed how you can make nonlinearity work to your advantage to learn a disproportionate amount from what you have. Time is also your ally, but the way nonlinearities help you manage that time may be a bit more counterintuitive.

As far as time is concerned, nonlinearity means there is no relationship between your rushing to market and growing fast. Similarly, when adoption kicks in, keeping up with demand will become the challenge. In the epilogue to this book, I discuss how this nonlinearity relates to technology adoption curves and to concepts

Time is also your ally.

such as the "chasm." In practice, nonlinearity implies that questions beginning with "When should I" or "When do I" (e.g., "When should I search for funding?" or "When do I know I have to pivot?") may make sense only when you retrace the steps of your story in hindsight. In foresight, they are meaningless. Because there is no linear relationship between time and progress, time is as bad an indicator of progress in innovating as is weather. And you can really only answer these questions with the Zen-like "when the time is right."

In foresight, the solution is to ignore time altogether and focus instead on preparing for scale. The time will be right when you are ready to learn from resources at the next scale in much the same way you learn from resources at the current scale.

Preparing for scale helps you master time in two ways: through resources and by sampling.

On the one hand, the strategy I propose to use resources at scale helps you manage time by increasing the breadth and rate at which you explore the problem. You can halt your exploration at any time and pass your understanding of the problem to future-you or to someone else. (See chapter 6.) Front-loading your exploration, when time is less costly, allows you to explore an entire space of opportunity. (See chapter 7.) At this scale, it takes weeks, not months, to explore a problem meaningfully. (See chapter 11.) And there is a simple way to make your project count toward your larger innovating objectives through documentation regardless of whether you've already found a way to bring it to the next scale. (See chapter 12.)

Nothing is lost if you stop, and you need no more than a month to explore at a small scale before it becomes advisable to distance yourself from the project.

On the other hand, if you choose to continue, you need to translate all the "When should I" and "When do I" questions into "How often will I" questions. For instance, fundraising is a *recurring* task that you will do every so often until your organization evolves to sustain itself through sales (a *continuous* task). At a high level, this simply means that as you progress in scale you will have to organize tasks by at least two criteria relative to time: tasks you will need to execute continuously and tasks you will need to revisit continually. In a way, the need to get organized emerges first from accepting that innovating is highly nonlinear.

In chapters 7–10, I discuss the dynamics imposed on your innovating by nonlinearity and reveal how to relate the skills you build through innovating to organizational building.

Takeaways

- There is a way to turn nonlinearity to your advantage and explore a space of opportunity systematically, rather than constraining yourself to a pre-specified future. It comes to you from chapters 3, 4, and 5:

  > Combine a few parts.

  > Start substituting those parts.

  > Talk with a few people.

  > Look for patterns in the combination of parts and insights from other people.

  > Outline a few (say, three to five) distinct opportunities for impact.

- It's great if you have an intuition for how to combine parts, but no intuition is needed. What matters is what you think is possible after the parts are combined. Learn what happens then. Ask yourself what new problem those parts might solve together. Pay attention to how people "misinterpret" your idea; those are small changes. Work the small changes back as parts: How are your idea and their "misinterpretation" the same?

- As your idea grows, you'll have more than parts and people; you'll also have sets of comments and ensembles of parts that you believe may work together in multiple scenarios. You may even be able to imagine going from one scenario to the next with only small changes to parts. Note that all this makes the concept of a near miss paramount.

- As your exploration continues, you should be able to demonstrate each opportunity with a different combination of parts. Each demonstration should be reasonably robust to slight changes in several of those parts. You have created a space of opportunity.

- When you look back after you're done executing your plan, you'll probably notice that you have used a lot of the insights that emerged from how you built this space of opportunity. To people unfamiliar with how you envisioned your opportunity, it will look as if you changed your idea continually, but that's just a by-product of how you built the connections between parts and impact. There was no need for you to figure out how to "pivot."

- Time is your ally.

Combine parts and insights together.

Learn a disproportionate amount from the resources you have.

Explore a large space.

Add, remove, substitute parts and insights.

Time is your ally.

Look for patterns in the

Rushing does not get you there sooner.

NONLINEARITY IS YOUR ALLY

What you know or have is an asset, not a liability.
Allow your innovating to venture into the impossible.

Progress is learning. Organization yields scale.

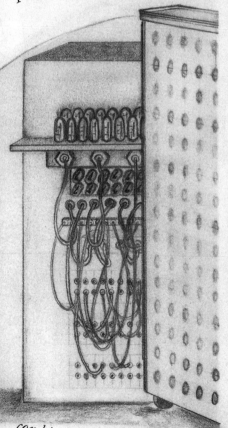

combination of parts and insights.

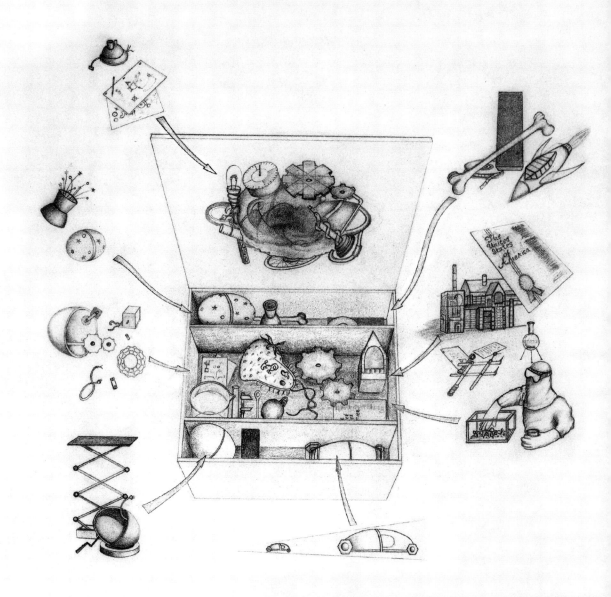

6

A KIT TO DRIVE
INNOVATION, ANYWHERE

There is a way to capture everything discussed in chapters 1, 2, 3, and 4 into a kit for innovating.

The very idea that you can somehow innovate using a kit probably sounds like nonsense at first, but it should make perfect sense if you think about what you actually have every time you restate your problem. After all, you can get started with parts lying around, and all you need is a hunch and a desire to refine a problem as you learn about that problem using those parts along with people. Your sense of impact drives you to make the problem tangible, and in the course of the process you discover and select from among multiple versions of the problem; you do so by outlining the next set of parts and people you need. Why, then, can't you take all of that, package it together, and pass it on to others to get them started with their own innovating? That's your kit. And because those others have their own sense of impact, you have every reason to expect that they will use the kit and end up addressing a different variant of the original problem.

So at a high level, the kit is just that: what you end up with after you try to restate your problem. It's your best understanding of the problem, the novelty, and the impact as it emerges from the parts and people you've brought together thus far. Still, the notion that you can produce a kit for innovating may seem like nonsense.

This chapter could easily have ended here with this closing: get a hunch; follow what you learned in chapters 2, 3, and 4; inventory what you end up with; restate the problem; and pass all of that on to someone else—or to future-you. And perhaps that's all you need to dive into the next chapter. That may be the case, for instance, if you are engaged in innovating right now—and if you are, feel free to skip to the takeaways of this chapter and then move on to chapter 7.

> You can tame any real-world problem with a kit.

But the entire book could also have started with this chapter. For instance, you may be interested in several other contexts in which you may find an innovation prototyping kit useful. You may need to set up an innovation process in your organization. You may find yourself needing to pass the results of your innovating to someone else so that person can use everything you've learned in his or her own innovating—something that happens a lot in any organization or business with a research arm. Perhaps you have generated some intellectual property and you want to engage others in devising ways to bring it to impact. You may be interested in technology transfer. Perhaps you want to teach innovating in a hands-on way.

Before I go into the details of how to build an innovation prototyping kit that helps you in each of these contexts, let me show you how it can be useful to get others innovating.

Innovating does not always begin with an individual. Even when it does, it is only a matter of time before a group of people come together. Sometimes, groups come together to engage in innovating from the very beginning. They become a team some time further down the line, when the group learns to communicate effectively around what amounts to a hunch, and when each team member has found a way to put to use his or her diverse background and interests to contribute to a shared objective for which they all feel ownership. Yes, that's right: a group of n people only does not a team make.

You may find yourself needing to drive one such group engaged in innovating, or you may be a manager in circumstances where groups of co-workers are expected to engage in innovation activities as a team. Perhaps you're expected to teach students about innovation as a concept and then give them something to do. What approach should you take?

In my experience, you have two general options. One is to drive the group to produce and accrue as many ideas as possible, usually in a vacuum, and then compromise to pick one. The other is to drive a team to work on making tangible a problem or problems given to the team at the start. The latter option replicates what Henry Ford, Theodore Maiman, Johnny Chung Lee, and the founders of Greenpeace all did. They began with a problem, and went from there. That's where a kit comes in.

The two options—idea harvesting and working with a kit—represent two very different paths on which a team will travel, and lead to very different outcomes. One leads to a list of idea statements; the other leads to

When a group eventually comes together, it isn't yet a team. Those people must learn to communicate about a hunch.

a set of strategic decisions to be made about a tangible problem. They also represent two fundamentally different belief systems about what an idea means: the former needs good ideas to be recognizable, whereas the latter presumes that what makes an idea good is that people can evolve it into something better. So, when making a choice you need to be cognizant of what you're deciding. Of course, if the first four chapters of this book tell you anything, it's that I advocate for taking the latter path of driving a team to work on a problem, prototyping the problem, and going from there.

As a matter of fact, I find the high hopes associated with idea-harvesting to be perplexing—whether it is in the form of a single brainstorming session or in the many forms a "call for ideas" takes in classrooms and corporations. This is worth a brief digression. Let me outline the problem I see in processes that begin with a "call for ideas." (The following diagram shows what I mean by the idea-harvesting path in a nutshell.)

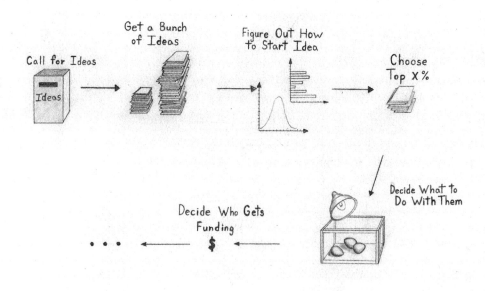

The Idea Harvesting Sequence

If you opt for accruing as many ideas as possible, that's what you'll get—a lot of statements of ideas. You now need someone to tell you which idea is "the one." By construction, that person is not invested in or tasked with advancing any of those ideas. There is no guarantee that even a single idea is actually good. As a matter of fact, the account in chapter 1 of how several innovations emerged shows that good and bad ideas are likely to be indistinguishable at this stage. Still, you need to review them all. You will be under pressure to pick at least one.

Suppose that I'm wrong and there are several dozen good ideas—by some absolute criteria of "goodness." Most ideas will be somehow inspired by what your organization does today, but it would be serendipitous if any one of them were to match a specific problem relevant to your organization's strategy or a real-world problem you're prepared to recognize.

The "Law of Large Numbers" applies here—in a negative way for innovation. The law expresses the idea that after you collect a handful of samples the addition of another sample is not going to change the mean significantly. Once you get a sufficient number of samples (in our case, ideas), you'll begin to see a convergence to the mean. In other words, the more ideas you have, the easier it becomes to gravitate to the "average" idea. Mathematics is clear on this: Unless you look at each idea for what it is, the only purpose of gathering many ideas is to assess some statistic—in modern parlance, we might call it "analytics," "idea funneling," or "dealing with idea overload." It all boils down to whether you believe that *the* idea is *the one* that enough people will have. You're taking it on faith that the "average" will be a "bright" idea.

The only thing we can say for certain is that your innovation process will begin each and every time with more (often many more) ideas than it ends up with—a funnel! But even this observation is marred by the paradoxes associated with hindsight—just like identifying something as an "innovation" long after the fact. Any selection process, good or bad, will act as a funnel. Naturally, the "good" selection process will be the one that leads to the "good" idea.

A Mathematical Nemesis of Idea Harvesting

The idea harvesting approach has another mathematical nemesis: the central limit theorem. If you use a bunch of random generation processes to amass ideas, the resulting means from all of them together will end up as a normal distribution. This will make it easy to characterize the distribution of ideas you harvested, but it will create the same imperative as with the Law of Large Numbers: taking it on faith that the "average" will be a "bright" idea.

As it turns out, mathematics comes with a rather large supply of practical nemeses that challenge the implementation of idea harvesting. There's even the infinite monkey theorem: that a monkey hitting keys at random for an infinite amount of time will almost surely type all the works of William Shakespeare. Similarly, a group of good innovators tasked with generating as many ideas as possible will almost surely—given sufficient time—invent everything there is to invent.

The only apparent way around the nemeses mathematics presents for the idea harvesting game is to employ some criteria to sort and filter those "idea statements." The objective is to define rankings and filters that will distinguish that one "good" idea from all the others. You can use several modern techniques conceived to address this kind of idea and information overload. However, they are all based on accruing first some kind of statistical understanding of the distribution of ideas, which brings you back to square one: You thought you were sorting through ideas to avoid the mathematical nemeses behind idea harvesting and the central limit theorem hit you while you were distracted in your tallying.

There are some good things about this. One is that, no matter the innovation process, you don't need more than a handful of idea-producing units to be able to characterize your innovation activity with the simplicity of the normal distribution. Another is that, given sufficient time, you'll invent something very good—and even more. In either case, you will need to worry only about training good idea scouts. By analogy with the infinite monkey theorem, that's a well-nourished team of expert readers who can read everything your monkey-innovators type and then tease out the works of Shakespeare as they emerge piecemeal and disorderly.

In summary, opt for this path and you may find that you've actually made things unnecessarily difficult. You began with a desire to innovate. You chose a process that moved you quickly away from innovating and from the idea generators and to selecting from an enormously large pool of potentially very bad ideas. Worse, there is no way to know how to make the right choice or even whether there *is* a right choice, and meanwhile the idea producer who could explain what he or she meant has long since "abandoned" the process. A lot of time spent, and none of it on improving the "idea." All you have is the conviction that among so many ideas one ought to be good. But even that belief is questionable. Applied to mining, the same deeply held belief might push you to discard prospecting altogether and instead acquire land, as if owning more land automatically increased the odds of finding the ore you want.[1]

You have translated a potentially creative endeavor into an actuarial one. And like a good actuary, your concern will be about a process that is efficient rather than a process that unearths what's best about an idea—a concept fundamentally different than unearthing the "best" idea. One thing is true, though: When it's your turn to report you'll be able to share lots of numbers—for example,

> We got x ideas. We developed a process A to trim them down by 50 percent, and then we sent comments back. Those ideas were further down-selected to a fourth of the original list, and with those we did a workshop with middle management, using Post-its, in which we discussed and re-ranked the ideas and ended up with another reduction by 50 percent. All in all, we engaged $x*2$ people from 50 departments throughout the company. Today, we're presenting the 'winning' 1 percent, and our leading idea is a fitness band that talks to your cell phone, pays like a credit card, and does laundry. What a great innovation!

A good group of senior managers would, and should, be horrified that so many person-hours were spent in an *actuarial* process initially conceived as a way to supply the firm with competitive advantage for years to come. Months were lost, lots of resources were used up, and the "idea" likely resembles something that already exists. All these outcomes are by-products of the process itself, the fatally flawed design of which ends up putting a halt to any work on the ideas before a selection is made. So, the ideas remain much as they were when the process began.

Let me motivate this last point, using the language I have been using all along: Idea-harvesting processes really set out to select among what I have called hunches. And because at first nothing about an innovation is really new, you should expect to find just that—a bunch of ill-formed ideas that are mostly wrong and ring as déjà-vu.

These idea-harvesting processes are appealing because they make accountability easier. There are lots of variants, many of which have tried to overcome the limitations inherent to the approach by incorporating ideas from the design world and from so-called "lean thinking." You can read about a representative example from Adobe in the box titled "Kickbox."

Good intentions notwithstanding, idea-harvesting processes probe creativity once and consume all resources as they sift through the mess that resulted from that lone outburst of unfettered creativity. But it takes more than a variant on idea harvesting to overcome a fundamental flaw in the approach: After the idea has been boxed, no one is actually working on it, yet you are still expending resources. Your innovation process begins with a waiting period.

Senior managers should be horrified by the resources spent on idea harvesting—an actuarial process— rather than on innovating.

Kickbox

Adobe's "Kickbox"—"open source" and available to anyone—is a good example of the approach to innovation that begins with accruing as many ideas as possible.* The small, red box is touted by at least one management professor as "containing everything an employee needs to generate, prototype, and test a new idea."** And what might *everything* be?

"When you break open the seal," the professor writes, "you'll find instruction cards, a pen, two Post-it note pads, two notebooks, a Starbucks gift card, a bar of chocolate and (most importantly) a $1,000 prepaid credit card. The card can be used on anything the employee would like or need without ever having to justify it or fill out an expense report."

In a video introducing the Kickbox, an Adobe executive characterizes it as "like the angel investor. There's everything inside you need to go do it, now."*** Of course, having an angel investor presumes you've already got an idea worth taking forward. In essence, the Kickbox and the brainstorming session both are begging for innovation in hindsight, pushing you to an endpoint without having made any determination of whether your idea needs some refinement as it reaches its future.

*http://kickbox.adobe.com/what-is-kickbox

**David Burkus, "Inside Adobe's innovation kit," *Harvard Business Review*, February 23, 2015.

***http://www.youtube.com/watch?v=VU4XYGuUh_o

Months of innovators-in-waiting and hunches in the dry dock follow the call for ideas. Some are waiting for selection to finish to approve a budget and others are waiting for approval of an idea. It isn't even clear that the idea supply and idea demand are all that well aligned with expectations about what will happen when the waiting is over.

There is definitely something paradoxical about ideating a process that begins with a call for ideas and moves soon thereafter into boxing ideas for sorting and voting and away from working on the ideas themselves. As I mention in chapter 1, coming up with an earth-shattering idea can easily turn into a rather stressful endeavor that postpones the kind of work that might help evolve an initial hunch—however ill-informed—into something valuable.

There is, however, good news. The same resources expended selecting could go to advising people to hold off on the Post-it and instead work a bit on making their problem tangible, perhaps all the way to a proposal. Whether you are tasked with implementing a new process to cull the creativity of your organization or are asked to respond to a call for ideas, getting tangible is so easy that you can always opt to go that way.

The concept of an innovation prototyping kit can help you with that.

This is how you go about it. As an innovator, your objective is not to share an idea right away but to evolve one until you feel it is worth diving into. Your ultimate goal isn't approval or selection of the idea; it is to advocate for a decision about a specific allocation of resources to bring your innovating to the next scale. From a management standpoint, your objective is to establish a process by which the best ideas self-select, so you minimize the guesswork regarding which idea is best and instead set yourself up to make decisions about how to allocate resources strategically.

In the following illustration, you see that the innovation prototyping path makes things easier. The idea is to entice a group to replicate the unassuming approach to innovating described in earlier chapters by driving a team to work on making a problem or problems tangible.

The actual setup will vary depending on whether you give these problems to a team at the start, whether you are trying to develop your own team, or whether you ask people in your organization to put forth ideas that you will then help evolve. But the mechanism is rather simple: conceiving an innovation prototyping kit to jump-start your innovating. And the objective is the same in all cases: to apply the same resources (labor and capital) you might otherwise have applied to combing through idea statements to help "hunches" evolve.

For the innovators-in-the-making you are trying to inspire, this is a challenge and an opportunity: You get to self-select by demonstrating the value behind your ideas, at the same time making your ideas better. For the managers and decision makers who oversee the process, the job is arguably closer to

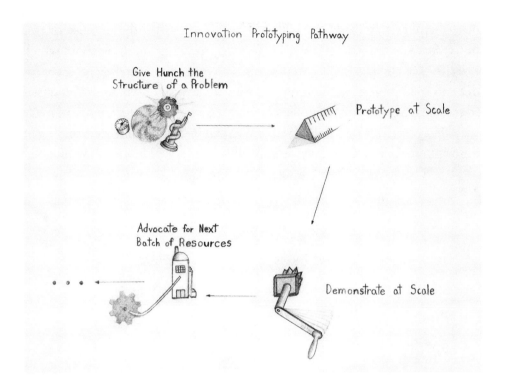

Innovation Prototyping Pathway

Give Hunch the
Structure of a Problem

Prototype at Scale

Advocate for Next
Batch of Resources

Demonstrate at Scale

The Innovation Prototyping sequence.

their current job descriptions: you can focus on mentoring in the process and making decisions about allocating resources, which is arguably easier and less stressful than forecasting which hunches are the wheat, which are the chaff, and how to separate the two.

Success is not measured by the number of ideas. It is measured by how much you've learned about new potential *outcomes* for innovating and by how much leverage you stand to obtain from spending additional resources to allow some of these ideas to progress. Ideas progress when they prove—or disprove—the viability of the concept at the next scale. Because the method is premised on the idea that everything can and should be demonstrated first at scale, thriftiness is already embedded in the process. Your job is to decide whether you agree that the "proofs" you'll get upon deploying resources in the manner proposed are interesting to you, or whether you believe some other "proof of concept" is needed. This all builds on the premise of this book: that you are best served when people produce something about which a decision can be made early on—whether that decision is to keep going, produce a variant, pursue something else, scale up, invest, or whatever—and that it does not take millions of dollars to arrive at a point at which you can make that kind of a determination based on fact; all it takes is an understanding of innovating as a scale-up sequence from which the actual innovations emerge.

What to do next may include assembling some different people, provid-

> You are best served by ideas that evolve into something that demands a decision: Keep going? Produce a variant? Pursue something else? Scale up? Invest? Whatever?

ing some funding, acquiring new or different parts, and so on—in effect, everything I discuss in this book that you can do to make a problem increasingly tangible. Next steps will be based on a decision about something real, not on an actuarial sorting process. Further, working with a team on this path keeps you firmly in the capacity of a *doer* involved in actual *doing*.

Along this path, your team confronts a problem that needs to be solved. The team makes the problem tangible, prototypes the problem at a small scale, and in doing so shows what needs to be done next. Then, if the time is right, a straightforward decision can be made: Should group X be given resources Y to accomplish objective Z? The senior managers who earlier were horrified by what was expended on an actuarial process ought to welcome this alternative.

Ultimately, however, it all boils down to your belief system about innovation. This book is based on the notion that simply accruing a bunch of ideas and then sharing them has much less value than *working on* ideas until the need for a decision is apparent. That decision may now be based on the intrinsic merit of what is being proposed rather than the relative merit of one thing against a set of ideas based on some arbitrary set of criteria (e.g., novelty, disruptive, creative, user-based, and so on). This matters because the "best idea" is actually a function of the endpoint. Early on, you're essentially guessing, and typically the criteria for selecting are just means to conceal that that's what you are doing.

For innovating, it doesn't matter whether there's a funnel paring down from a larger to a smaller number of ideas, or whether you begin with one idea and stick with it. Creativity, novelty, and the disruptive nature of your idea are all outcomes. What matters is *impact*. From the outset, an innovation prototyping kit helps you place the emphasis on the problem, make a tangible demonstration, and proceed to reduce the process to "decidability"—meaning that "ideators" can advocate for a decision based on real next steps.

Whether you are setting the process, responding to a call for ideas, or simply creating a team, you need a way to embody the problem you are solving so that others can join you in the endeavor (as teammates, directors, funders, etc.). In the next paragraphs, I build on chapters 1, 2, 3, and 4 to describe what actually comprises an innovation prototyping kit and how it embodies the essence of this approach to innovating—namely, parts, scale, and people—and the logic of problem solving. I also discuss how to set up a process that helps teams engage in innovating, and what to expect from a team when its turn comes to advocate for resources it needs next.

What is an innovation prototyping kit?

In essence, an innovation prototyping kit brings together a problem, parts, impact, and people—the components discussed in the preceding chapters. It is used to direct people toward innovating and solving a real-world problem. The basic idea is that one such kit contains all you need to play with a problem at a table scale. But it is otherwise open-ended.

The very idea of such a kit is premised on some notions discussed in chapters 1 and 2. The first is that to start innovating, all you need is a hunch. In other words, you don't need an earth-shattering idea. You don't even need to confront a big problem, only a real problem—and have only a hunch about it. Having a hunch, as I write in chapter 2, is the most uncomplicated starting point for an innovation—one from which you can set forth repeatedly without fear of being wrong. A hunch, when phrased as a problem, points to a "family" of problems to be discovered and potentially addressed. Thus, a problem—or, more accurately, a hunch with the structure of a problem—is the first component of an innovation prototyping kit.

The second component of an innovation prototyping kit is parts. Basically, a kit need contain only a set of mostly accessible parts, along with some pointers regarding how others have achieved impact addressing adjacent problems using the same parts or similar parts. What is accessible changes from one context to the next. If you are working within your own means, what's accessible is probably anything you have on hand, can pick up at a local store, or can order online at a reasonable cost. This is in a sense the latter-day equivalent of Dr. Maiman ordering a flash lamp from a photography catalog. The Internet is your catalog, and you may even qualify for two-day shipping. If you are working in an organization, what's accessible expands. In principle, anything your organization does, procures, or manufactures today may become the basis for a part. That is how you may get your employees to innovate on the basis of what you already have, and to arrive at a radically new solution. "How others have achieved impact" can be easily described with a few examples of successful organizations. The purpose of such examples is to reveal the diversity of ways in which people achieve impact with parts not unlike those being proposed, and to provide a first indication about the kinds of expertise you may need to look for—that is, people.

The third component of an innovation prototyping kit is people. That includes the group of people you've brought together to work with the kit—the team—if that's what you are doing. It may also include a list of people who have agreed to make themselves available to a team, such as mentors and subject-matter experts inside or outside your organization. If you decided to build a kit out of a technology born in your research organization, the list may include the inventors. The list can also be simpler, limited to an enumeration of your best guesses of the profiles of people the innovators may need to find.

The final component of the innovation prototyping kit is what I call a "primer," and it is conceived to help a team overcome the fear of being wrong or of not knowing enough about a topic, the parts, or the modes of impact. This is the same fear that comes up in chapter 3, "a fear about parts and about getting tangible too quickly." And it emerges from a confusion: adults who are too afraid of being wrong think they ought to know a lot about the parts before they use them. For instance, you may think you need to know a lot about electronics to use a micro-controller, or that you need to know a lot about marketing to make a meaningful assessment of user profiles and preferences. The truth is that such knowledge may help you, but simply isn't necessary. Your purpose is not to advance the state of the art in either electronics or marketing, but rather to use parts that may be common to one of those disciplines for the purpose of advancing your understanding of the problem. You are an experimentalist. You just need to know how to put them to use safely and how to interpret the results they produce. That's a much more accessible goal.

As you evolve your understanding of the problem, the need to fine-tune some aspect of your project or make specific modifications to some parts will lead you to search for people who have specific expertise you need. The primer is an instrument to help a team get up to speed with how to use the parts you assembled in your kit.

This primer fulfills an important objective. Most treatises on innovation and entrepreneurship encourage readers to avoid getting too deep into technical experimentation before understanding their users. In fact, they are asked to avoid it altogether. That's a bit extreme. You can as easily blindside your *own inquiry* by obsessing about "users." Getting technical too quickly in, say, chemical engineering is as bad as getting technical too quickly in marketing. You're equally likely to be wrong about the technology, the users, and the "product."

Of course, there is nothing inherently bad in being wrong—provided you set yourself up to discover that you're wrong before you're knee deep in it. The primer is meant to help you be wrong quicker. It teaches you how to use whatever parts you have or obtain to materialize ideas, and how to approach people to extract information for innovation prototyping. It helps you connect both parts and impact at scale so you get only as deep as the current scale of your project warrants.

The primer helps innovators focus on using the functionality the parts provide to demonstrate their own version of the problem at scale. You can achieve that by coupling very basic knowledge of how to use the parts themselves with some guidance on how to view those parts as scaled-down representations of a larger reality—not unlike the examples in chapter 3.

For instance, you can create a scaled-down version of the signage in Times Square. Simply use a simple circle circuit with an LED connected to a microcontroller that turns the LED on and off with some periodicity. Instructions for how to build such a contraption are available for free online. If you train yourself to focus on what the instructions say needs to be done rather than on the accompanying technical explanation, you can get your system working in about half an hour, even with no prior knowledge of electronics.

In one day, with parts available online, you can build the image of your business online, including a purchase order system.

You can use your own everyday experience as a supermarket shopper to explore how users influence supply chain management. Suppose your favorite brand of coffee is always undersupplied. Instead of buying it "just in time," make a point to stock up; every time you see it in the supermarket, empty the shelf. A few weeks or months later, depending on how inventory is managed and analyzed at your local supermarket, you'll see the supply increase—and you will have an opportunity to start managing your own stock.

No matter the topic, there is a way to bring knowledge down to a scale at which you can try it out and relate to it through impact, which requires little or no experience. That's how you set yourself to learn incrementally about the ways in which a specific body of knowledge or practice applies to your problem. And that's also how you can arrive quickly at a tangible demonstration of a problem. The primer ought to help your innovators get started doing just that, with the parts you provide.

There is an advantage to seeing the parts you are about to assemble by how they represent a larger reality rather than by how they work. Your trial and error will benefit from all the nonlinearities of innovation. Let me explain first by example. In the Times Square example above, if you replace the LED with a buzzer (a task that requires a small investment of about ten seconds of labor and a few cents of capital), the same circuit becomes a scale reproduction of an alarm clock. A small change like that in how you use the parts may translate into radically different opportunities for impact. The converse is also true: Seemingly innocuous changes in the scope of the impact you seek—such as the insights you may glean from speaking to someone—may fundamentally change how you assemble parts and even which parts to bring in. Fortunately for you, the emphasis on first demonstrating things at scale makes that kind of trial and error inexpensive and fun, no matter how "dramatic" the changes may be.

It may seem ludicrous to think that small changes can have such an enormous impact. It defies the habit of thinking linearly. But you ought to hope it's true; otherwise, you would have to count on dramatic and disruptive change being the only source for equally dramatic changes—an idea that seems terrifying. In chapter 5, I note that it is the nonlinear nature of innovation that creates this benefit from small changes. There, I use the vacuum tube as an illustration, noting that electronics emerged from combining two well-understood parts, a light bulb and a thin metal plate. All it seems to have taken was for Thomas Edison to accept changing parts as an operating principle and for Sir Jonathan Fleming to realize later that this new combination could be used for something else.

All I am suggesting is that you can use a primer to help your innovators think this way and, in doing so, use nonlinearity to their advantage. If you have not already blindsided your own inquiry with "user" or "technology," you stand to benefit a lot from these nonlinearities and you will be able to evolve the problem that gives you purpose continuously merely by replacing parts and impact and asking yourself three simple questions over and over:

- What large-scale problem are these parts evidence of?

- What extra evidence do I need to generate to bring that problem down to table scale?

- What parts do I need to bring together to generate that evidence?

This is the mind-set the primer, parts, impact, scale, and people ultimately enable. I conceive of the innovation prototyping kit as combining the simplicity of "tinkering," the ethic of "do-it-yourself" for hobbyists and others, and the familiarity and ease of use associated with the science kits many readers will remember from their youth. But there are very important differences, most notably that an innovation prototyping kit is meant to support open-ended exploration.

Trial and Error and the Power of an Open-Ended Kit

Innovation requires trial and error about a problem—indeed, to a degree significantly greater than afforded by the existing "tools" for innovation, be they patents, scientific papers, development kits, business modeling kits, or sets of tools for user discovery. To build on the premise that innovating is akin to learning, we need tools for innovation that aim at teaching something. Those are a special kind of tool because they must help you learn something about something no one else knows enough about. In other words, you need tools that empower learning and open-ended exploration.

Traditionally, kits have been used as a learning "tool." They have also been used as a development "tool." So it makes sense to think of *toolkits* that could somehow combine both purposes.

The use of kits as an educational tool to help people understand basic science has a long history.[*] Chemistry sets have been around since the eighteenth century in Germany and England, and in the early twentieth-century "toy" chemistry sets for home use were introduced, aimed mostly at teenagers. Science kits, though, are instruments with which to learn particular aspects of whatever is the model of science at the time the kit is created. But, as Thomas Kuhn implied in *The Structure of Scientific Revolutions*, science tends to rewrite itself every so often, so the model of science a kit is teaching may survive for only two or three generations.

The use of "development kits" helps people build from or interface with an existing product. Several software companies and several hardware companies use standard development kits to broaden the applicability of their products. Those kits, however, are only as good as the products upon which they stand.

The main difference between such educational and development kits and innovation prototyping kits is that the latter are meant to be open-ended, designed for trial and error and for experimentation. The aim is not to help you understand some particular underlying science, marketing, or technology, but rather to help you advance in your thinking about a problem—which means learning about the problem. There are no instruction sheets that lead you to a particular conclusion show you how to build a particular "thing." Nor are there "fake" open-ended questions—Socratic questions—at the end with suggestions about parameters to change or explore further. Rather, the problem being tackled always remains at the center of all activities with an innovation prototyping kit.

For a person leading a team, this means a kit can be used to center the attention of innovators on a set of resources and a specific family of problems. The creativity comes from how team members use their backgrounds, knowledge, skills, and abilities to communicate with teammates and others beyond the team to reduce the problem to tangibility.

In short, an innovation kit simply allows you to use "parts" and knowledge to arrive at novelty and impact through an open-ended process that is not aimed in the slightest at getting you to a particular conclusion.

[*]See Michael Schrage, "Kits and revolutions: An MIT economist's lesson in Kitonomics 101," http://mambohead.com/wp-content/uploads/2011/12/Make_Ultimate_Kit_Guide.pdf.

The innovation prototyping kit exists in the space of tools and even "movements" today that provide resources for individuals and organizations that promote knowledge self-sufficiency: the do-it-yourself and maker movements, the hackerspace movement, a variety of open-source platforms for technical and non-technical tinkerers, and so on. Increasingly, people are trying out new approaches to acquiring knowledge, distributing ideas, achieving impact, and leveraging new distribution channels to bring ideas and innovations to scale more efficiently and broadly than in traditional channels. The innovation kit is part of that world.

How to build an innovation prototyping kit

When you're done constructing your innovation prototyping kit, it will comprise a hunch dressed as a problem, a set of accessible parts, pointers to impact, pointers to people, and a primer on how to work on the parts and the impact at scale. In my experience, it's possible to build such a kit to seed the exploration of any kind of hunch with no more than a couple of pages of information and a few hundred dollars' worth of parts.

It begins with the process I outline in chapters 2, 3, and 4. Beginning with a hunch:

1. Consider what a solution must accomplish to solve the problem, and come up with a "recipe" that would allow you to verify that the problem is indeed solved. Try to resist the urge to come up with a solution at this point.

2. Write a general problem statement that brings the two together.

Note that, contrary to your first intuition, the kit will be stronger the more general your problem. So, while you need to be *precise* in your assertions, there is no need to be *accurate* at this stage.

3. Take a few seconds to imagine a few solutions that would, in your opinion, solve different versions of your problem. List them.

4. Break your hypothetical solutions into components.

5. For each of the parts, produce a brief explanation of how to use it and illustrate with examples of how each part models the kinds of impact you've articulated. Do the same for each kind of impact. Given the parts you've chosen, produce a brief explanation of how you would combine them to simulate each kind of impact at scale and what needs to be scaled up.

6. Assemble your kit.

Let me comment on these last four steps together so you can see how a kit emerges.

The main purpose of listing a few imagined solutions for different versions of your problem is to inform your quest for parts, people, and examples of impact. The solutions need not be earth-shattering; in fact, they can be trivial. Nor is it necessary to flesh them out completely.

The solutions you identify need not even be new. Draw inspiration from things already done. At this stage, a solution could be something like "adding Y to what company X already does." This list will also serve you as you seed the thinking of your innovators. It should inspire them to think creatively.

Having multiple imaginary solutions is most helpful—and the more different the imaginary solutions are, the better. It reinforces the open-ended nature of the task. Also, the more humorous and other worldly your "seeds," the easier it will be for your innovators to embrace "being wrong."

I find that using sci-fi books and movies to inspire some of these "seeds" tends to help achieve the objectives. Gene Roddenberry had flip phones in the original *Star Trek* series (mid 1960s), and characters in *Star Trek: The Next Generation* (late 1980s) carried around something that looked a lot like an iPad. In chapter 1, I mention Johnny Chung Lee finding inspiration in the movie *Minority Report*. When Greenpeace established itself as a non-profit organization it decided to include the term "Foundation" in its name to honor Isaac Asimov's *Foundation* trilogy. Humans were first sent to the moon aboard a bullet-shaped rocket, as Jules Verne had written more than a hundred years earlier in *From the Earth to the Moon*.

Breaking down your hypothetical solutions into components (step 4 above) involves asking a number of questions. Which parts would help you simulate one of those solutions at table scale? Do not limit yourself to the physical parts, such as a micro-controller, some piping, or plaster. For instance, if there is a regulatory component, how would you simulate that at table scale?

Given those components, whom would you ask for information about each?

What people or companies are currently producing similar solutions or addressing similar problems? Why? What makes those people or companies special given your problem?

Your answers to all the questions above tell you what parts, people, and examples of impact you will want to include in your kit. You will benefit most from being precise in answering the questions without worrying about being wrong—that is, you do not need to be accurate.

Producing the explanations called for in step 5 above keeps you focused on impact. Try to supply examples of use in which replacing one small part changes the mode of impact dramatically. For instance, if you are thinking about a combination of hardware and software for researchers, change the "user" part from "researcher" to "educator." The former leads you to a terminal market (research), whereas the latter guides you to a strategy to grow markets through academia that has been used by companies such as Math-Works, Apple, and Facebook. You just changed one part.

When you're ready to assemble your complete kit, it will include the following:

- a general problem statement (from steps 1 and 2)

- a set of quick ideas innovators may see as examples of things that could lead to solutions (this can be inspired by step 3)

- a set of parts, examples of impact, and a list of people or profiles of people who may know more about either (from step 4)

- a primer that illustrates how these parts may be used as illustrations of larger realities and how to break larger realities down for simulation with the parts you have (from step 5).

Just as the solutions in step 3 need not be earth-shattering, nothing about the kit is meant to be earth-shattering. You should view it as a beginning, a place to start. Much like your first pass at defining your problem, discussed in chapter 2, the kit is meant to provide some clarity regarding what your problem actually is. You are just giving a loose sense of direction to your innovators and inviting them to imagine—something they will have to do often as they set out to view how each part represents a larger reality.

If this all feels too oversimplified, it's because it is indeed quite simple. The rest—what people praise as creativity, imagination, the "eureka moment"—comes later, the result of playing with parts and impact to tame an imaginary problem into becoming real. Your innovators will provide that as they apply their creativity continuously to evolve the problem, empowered by a kit and enabled by the process you choose to implement. Or they won't. One thing is certain: The kit will not *create* on its own.

Setting up a process for people to use innovation prototyping kits

When you bring a number of people around a hunch in this manner, the destination that gets shaped is a function of who they are and how their interests overlap—provided the problem is open-ended. As the problem evolves, part of the job of the team using the kit is to move from that initial list of parts and people to one with new parts and new people whose input is most relevant to the evolving version of the problem. The challenge is to conceive of parts with which to demonstrate the problem that are commensurate with how well the people involved understand the problem today.

Three principles to follow: open-ended exploration, evolution through the combination of accessible parts and people, and challenging people to put the demonstration of value ahead of requesting sizable amounts of money.

There are three principles your process must follow: open-ended exploration, evolution through the combination of accessible parts and people, and challenging people to put the demonstration of value ahead of requesting sizable amounts of money. Your job is to set up a process for innovating that enables that kind of inexpensive exploration, so that resources go to empowering the creativity that emerges from recombining parts for the purpose of defining a tangible compelling destination.

As your innovators evolve the problem, so too does the level of sophistication with which parts and people need to come together for impact. Soon they will need to think about an organization that solves a problem. The organization will need a specific set of resources to get started and to eventually solve the problem sustainably and systematically. When the team members get there they will find themselves needing to advance their problem to a larger scale, and you will be ready to discuss next steps and value creation milestones.

That's the outcome you seek. When teams are ready to advance to the next scale, they ought to be able to articulate a problem an organization can solve. They will be able to outline the way an organization can be put together incrementally to solve that problem, the resources needed, and the impact that ought to be expected from engaging in such an endeavor. By then, their prototype will outline an organization, including where innovation is needed. The prototype of the organization ought to make it easier to imagine the scale-up needs and to assess the risks and uncertainties that remain given their proposed plan of action.

I have implemented this approach successfully in classes and workshops. As long as you set a limit on the time that should be spent on any one innovating project before reaching a decision on a destination and what needs to happen at the next scale-up stage, the process translates naturally to many other contexts. I discuss this in more detail in chapter 11.

When to consider producing an innovation prototyping kit

Working in just about any profession has a ready-made aspect. If you tell me you're a chemist, I will imagine some tools of the trade such as beakers and solutions and Bunsen burners. If you tell me you're an electrical engineer, I will imagine circuits, semiconductors, and oscilloscopes. If you tell me you're a marketer, I will imagine messaging, product placement, and focus groups. Tell me you're a lawyer and I will imagine contracts, negotiations, and precedents.

So, I am an innovator. What do you imagine? If you imagine product placement and user surveys, you have me confused with a marketer. If you imagine beakers and solutions, you have me confused with a scientist.

As an innovator, the tools of your trade are problems, parts, and people. Sure, the parts may be shared with one of the professions above, but for you they derive purpose from the problem, not from some principle or endpoint presumed by the profession. The "kit" for an innovator ought to be an instrument for people to "use" parts and knowledge to arrive at novelty and impact.

People often associate innovation with patents and research papers. But patents as currently constituted are hardly a way to engage in trial and error as a path to innovation. They are legal tools that gain their full strength from their defensibility in court. Research papers are scientific tools that report findings that advance the state of the art and gain their full strength from their reproducibility. Much of the same could be said about scientific kits, development kits, product ideas, and other "tools" often associated with innovation. They all need translation to become usable for innovation.

Translation

There are several contexts in which an innovation prototyping kit offers a useful way forward.

Some organizations already produce development kits for their users to build upon their products. Those kits tend to be focused around existing products, but they too may become a part in an innovation prototyping kit. All that is needed is to augment the development kit with additional parts and components selected to address one or a family of problems. As a matter of fact, this is an indirect way to innovate upon existing products—viewing them as parts of a larger system and, as such, subject to further evolution.

In an academic environment, a common problem is moving research beyond scientific papers—to the next graduate student or to society. There is a lot of information that never makes it into *the* paper, and to an aspiring entrepreneur what's needed is not the specifics of the results section and the challenge to reproduce those results but getting access to what the researchers learned along the way. An innovation prototyping kit may help overcome that obstacle. In the language of this book, the problems the authors of a paper spelled out in their grant application are hunches, the research they conducted revealed parts they used to formulate that problem tangibly, and the results they obtained were new parts. Everything they learned could be encoded in a primer as variations one might introduce in the parts to explore different avenues of impact.

Finally, you could create innovation prototyping kits from intellectual property such as patents. At a high level, you may view patents as a legal instrument to enable transfer of knowledge about new inventions to society—so society can build upon what's discovered. Understood in that way, patents are legal mechanisms that enable innovation while rewarding inventors for sharing their knowledge with society. But they are just that: legal instruments, contracts with society, as well as instruments in litigation. As such, they are often criticized for stifling innovation. I choose to believe that if patents were easier to engage with it would be easier to appreciate the many ways in which several patents enable new inventions and applications that contribute to "increase the pie." Innovation prototyping kits may become such vehicles for experimentation.

I believe that innovation requires significant trial-and-error work on a problem—a kind of trial and error not afforded by patents or scientific papers or existing kits but perhaps afforded by innovation prototyping kits.

The logic of trial-and-error exploration that makes kits work defies linear thinking. The alternative— that disruption comes only from dramatic changes —is terrifying.

Takeaways

• You could set out to harvest ideas, and thus postpone innovating, or you could set up a way for *anyone* to start innovating right away. That's your choice.

• Professional innovators need their own tools of the trade to operate on problems by working with parts and people. An innovation prototyping kit is one such tool. It's easy to build. You can build one out starting from anything—for example, a product idea, a hunch about a problem, a development kit, a research paper, or a patent.

• In essence, an innovation prototyping kit is an invitation to accept that innovations are not planned but rather emerge as you tackle a problem. That is, the actual innovations emerge from your efforts to combine parts and reasonably accessible knowledge as you try and err about a problem.

• You can get parts from a catalog or simulate them somehow; the knowledge about those parts or about the problem you do not already possess resides in other people. The problem that gives you purpose is a problem only if it has not been solved, so you ought to accept the fact that, as you go about understanding that problem, you'll come up with something new.

• You need something to empower your exploration, and it ought to consist of things that already exist. If you accept that the trial and error reveals the innovations needed, and that you can at the very least simulate their impact using other parts, there is no point in trying to plan a "great innovation" or culling innovation ideas from the get go. You can choose to focus on the problem and simply get started.

• An innovation prototyping kit packages a hunch about a problem and helps others to get started.

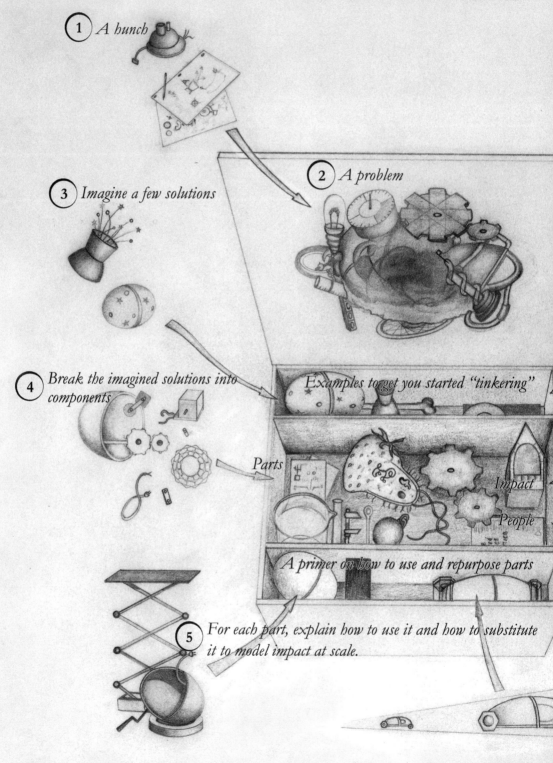

1 *A hunch*

3 *Imagine a few solutions*

2 *A problem*

4 *Break the imagined solutions into components*

Examples to get you started "tinkering"

Parts

Impact

People

A primer on how to use and repurpose parts

5 *For each part, explain how to use it and how to substitute it to model impact at scale.*

AN INNOVATION PROTOTYPING KIT

Inspiration from wherever to imagine solutions

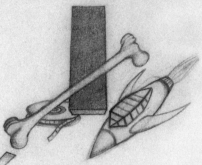

Inspiration from what you already know or have

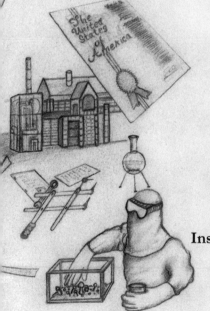

Inspiration from use

Sample Sources of inspiration

Another product
Science fiction
What you do or procure
Other companies as templates
Tools you already use
Knowledge you use
Existing business units,
R&D, patents, Academic Papers
People who use knowledge

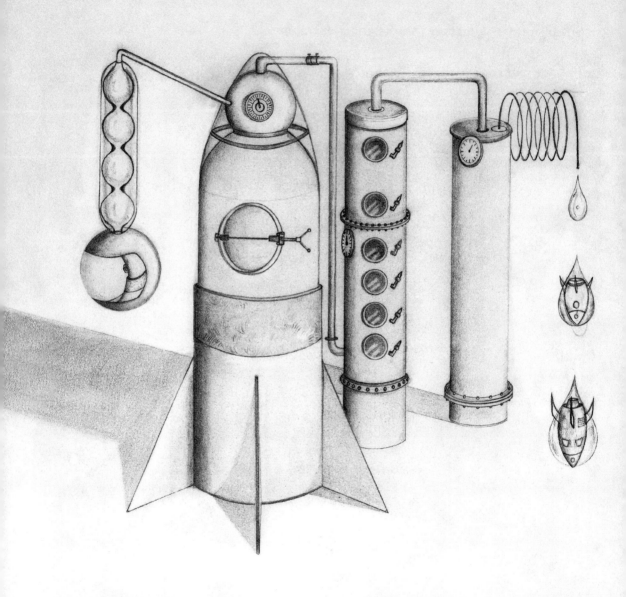

OPERATING ON THE PROBLEM
THROUGH TRIAL AND ERROR

"Failing" sooner buys you time. The further along you are in scaling up, the more attention—time and money—fixing a failure demands. Better to spend your time and money failing earlier, when you can apply what you learn to make the next scale work. You need to stay at least one step ahead of your failures. You do that by striving to fail first at scale at which failure is indistinguishable from an error.

This is not an argument to postpone scale-up, nor more praise for "fail fast, fail cheap." Rather, it is a reminder that mastering a problem takes practice and practicing implies making errors. You can let those errors find you when they *do* happen—and risk failing—or you can plan to make them, and learn.

So, you start by assuming you'll fail to solve the problem that gives you purpose.

That problem is wrong. Let's figure out how, so you can fix what's wrong. That's how you will solve *for* the problem and the problem itself. That's what I mean in chapter 2 when I write that innovating is akin to a general kind of problem solving in which you discover both the problem and the solution as part of the process.

As you amass parts and insights from people, and you change, add, repurpose, or discard them (as described in chapters 3, 4, 5, and 6) your objective is to make them work together and to investigate how your problem is the wrong version to solve—and so you can find a better one.

How could your problem be wrong? At "best," your problem lacks resolution; at "worst," your problem misses the point. The best way to find out is to try reproducing your problem. That's what I call solving *for* the problem.

What you are doing is reconfiguring your endeavor in a slightly different way. You are *de facto* building at one scale a proof of concept of what you will be able to accomplish at the next scale. When you land on the next scale, a proof of concept built in this way will buy you the time you need to address the new set of things that are only now apparent and that might make you fail at the next scale or the one after. This is how you layer proof of concept upon proof of concept.

"Failing" sooner buys you time and at scale is just an error.

Focusing on what might make you fail is not how we are typically told to think about achieving the super-glamorous-hipster-entrepreneurial future that will make you a cool and famous innovator. But it is a good way to set you on a path to find a set of solutions that have chances of scaling up and thus are worth pursuing.

As you aim to achieve impact with your innovating, you need to find the "tough" questions to ask about your problem that will help you fail at the lowest possible cost. The tough questions are the ones that help you see how

> Mastering a problem takes practice and practicing
> implies making errors. You can let those errors find
> you when they do happen—and risk failing—or you
> can plan to make them, and learn.

being wrong can help you progress through scale. It can be helpful to think of all the ways your endeavor might be killed.

Consider the approach astronauts use to prepare for a space mission. In an October 2013 interview with National Public Radio, Canadian astronaut Chris Hadfield described preparation for a mission to the International Space Station for an extended stay:

> Half of the risk of a six-month flight is in the first nine minutes, so as a crew, how do you stay focused? How do you not get paralyzed by the fear of it? The way we do it is to break down: What are the risks? And a nice way to keep reminding yourself is: What's the next thing that's going to kill me? And it might be five seconds away, it might be an inadvertent engine shutdown, or it might be staging of the solid rockets coming off. ... We don't just live with that, though. The thing that is really useful, I think out of all of this, is we dig into it so deeply and we look at, "OK, so this might kill us, this is something that would normally panic us, let's get ready, let's think about it." And we go into every excruciating detail of why that might affect what we're doing and what we can do to resolve it and have a plan, and be comfortable with it. ... [1]

The process Hadfield describes is a way to articulate, and gain a greater understanding of, just what's needed to survive. His straightforward question—"What's the next thing that's going to kill me?"—summarizes the relationship the astronaut has with the rocket, the mission, and life itself. It helps prepare the astronaut to overcome anything that might kill the mission. And it's relatively simple because it pushes the astronauts to search for certainty, which may come either in the form of a protocol of action or a request to change something in the rocket or the mission.

The approach Hadfield describes asks questions that have definite outcomes, and with which astronauts can begin to expand their understanding of the mission and the rocket. It might play out like this: If problem X manifests itself, we have 5 seconds to fix it before the rocket blows up. But we know we need 20 seconds to fix it, so if problem X happens, we die—unless we can somehow modify physics (i.e., engineer something) to give us the extra time or we find a way to fix it in less time.

It's morbid, to be sure, but also remarkably straightforward. And it informs the entire endeavor. Perhaps there's a relatively simple fix, such as adding a sensor or using different piping on the rocket. Perhaps the solution is more complex and requires a different engineering design.

This is actually about understanding the problem, not "solving" it per se—even if a solution eventually emerges. There is a problem: getting to the space station. There is a vehicle (literally) for solving the problem: a rocket. There is a plan of action: the prototype the astronauts question over and over, with their questions focused on what might lead to their deaths.

You can use the same approach Hadfield describes to fail at scale and avoid dying. If the event that would have otherwise killed the mission happens, you'll be prepared. You gain a full understanding of your problem and the way you think you will solve it. With that understanding, you can advance to each new scale. Resolve that if the overall mission *does* end up failing, it will be because of something you could not have known or predicted.

The non-astronaut in you may have applied a similar logic under much more mundane circumstances—say, planning a trip to a warmer destination. You may have created a checklist of everything you thought you needed to do or pack before leaving. Your checklist included a swimming suit, but because you wanted to travel light you wondered whether you should instead buy a swimming suit at your destination.

How did you decide whether to pack your swimming suit? You tried to figure out what would "kill" the trip. If you decided not to take one, and your hypothesis that swimming suits would be available to buy when you got there was wrong, your trip would be a disaster. So you tried to validate your hypothesis by asking friends who had been where you were going whether there is a local swimming suit store. But a lot could still go wrong. If the store is closed the day you arrive, you might lose a day of swimming; your size suit could be out of stock; maybe the styles and colors available are so odious to you that you couldn't bring yourself to wear one.

In the end, maybe you decided to pack your swimming suit after all, to avoid those possibilities that could "kill" your trip. Of course, you don't know whether it might rain the entire time you are there, ruling out swimming altogether, and in the end all your thinking may not have mattered. You throw your hands up, exasperated at the prospect of rain, and make your decision not to bring or buy a swimming suit on the basis of no information at all. That's your minimum viable "product"—not doing a thing.

On Externalities and Not Doing a Thing

An all too common way to misapply the logic of being wrong leads to rejecting perfectly valid ideas on the basis of having no information. My swimming example makes this apparent: you imagine rain, throw your hands up, and give up on swimming altogether.

It is preposterous to make a decision by relying only on statistics (say, about others' trips) or on a suspicion that something ominous could happen. Yet I see people apply this reasoning repeatedly.

The faulty logic goes more or less as follows: You imagine something is possible. You try to imagine things that might go wrong. You come up with one, claim victory, and decide that what you first imagined cannot be done. Thus, you have "proven" it was impossible. It's the perfect self-fulfilling prophecy: "I will not do it, and hence it does not get done."

You forgot you were supposed to find evidence. By instead letting everything happen in your head, you kept alive the myth of the impossibility of it all.

Aspiring innovators inadvertently use this faulty reasoning to discard perfectly valid ideas. After a while, proving "the impossibility of it all," they settle for ambitionless ideas that create the illusion of clarity because they are copycats of something someone else is already doing or because they are problems already solved. Or maybe they settle because they were simply tired of throwing their hands up over having to throw their hands up—and so they throw their hands up.

For example, people tasked with finding information from other people will persuade themselves no one is available to answer their questions, and so they don't even try. Later they argue that not trying is evidence there really was no one to be found. I congratulate them: truly the best way to persuade yourself something is impossible is to try really hard not to do anything.

Often, student teams come up with ideas to apply some new technology in, for example, a health care setting. Initial enthusiasm usually gives way to getting hung up on the belief that some regulation will make the idea too difficult or expensive to implement. Then they give up; a decision is made with no actual information. I explain to them that regulation might also make things easier, and that they will only know by trying to find out. After all, regulation provides the equivalent of a standard they don't have to develop on their own.

Back to the checklist for your trip, which is probably based on your experiences. Each trip you take informs your planning for the next one, helping you create a checklist of things to do, questions to ask, and things to pull together here and there. That checklist defines your problem indirectly by all the things you can imagine might go wrong and that you want to head off at the pass.

My examples of a mission to the International Space Station and traveling to a warmer destination illustrate two extreme scenarios in which you advance toward success by making it your purpose to negate that the idea might succeed. A third example will help show the mind-set underlying this method of interrogating an idea. Say you are about to embark on a trip to Brazil, your very first foreign trip. You can play the trip out in your head. You arrive at the airport in Rio de Janeiro, show your passport, and are informed that you also need a visa. Entry to Brazil is denied, you have to take a plane back to where you started from, and your trip is "killed."

If you had created a checklist of what you think you need to bring or do, you would add "visa" to it, next to "passport."

Now you're focusing on what might "kill" your trip. Perhaps your trip includes travel to a remote part of Amazonia. There, you may run out of *reals*, and discover that you are prohibitively far from anywhere you can get more cash. Since cash is vital to continuing on, your trip again is "killed." You'll probably add "extra cash" to the checklist; you might also add "short-term job" to your trip plan, too.

On each imaginary version of your trip, you learn something new that matters for success, and in so doing you create an increasingly more detailed checklist.

Your checklist moves you forward, but the only way to know if your checklist includes everything it needs to include is by moving backward—specifically, trying out a checklist, getting "killed," and going backward (in a sense) to create a new checklist based on your failure and what you learned from that failure. Eventually, the checklist has grown into a plan.

Surely your experience will change as you go, and unanticipated externalities will reveal themselves. The practice you acquired negating your trip will help you identify those externalities as things you could not have easily predicted when you were working at scale. It will also prepare you to learn from them.

As far as the problem you're working on is concerned, you need to make it a routine to ask questions that involve definite outcomes. Some of the answers will be things you know; other information you'll have to search for, and maybe ask others. The decision of what to do, however, is ultimately yours, as are the trip and the experience, and as would be the failures.

The Rationale for the Trip Examples

Using trips as the examples here conveys that what matters is experiencing a trip for the reality it offers, not for everything that could have easily been prevented. The reality that will shape your innovating lies ahead, and things can look quite different before and after your journey.

The exercise of negating your idea does more than help you anticipate disaster and prepare you to execute on a plan. It forces you to think about your idea in terms of strategies that increase your resilience and determination to get through the entire journey. When reality hits, you may not have anticipated some specific thing, but you will be more prepared to react.

As I wrote in chapters 1 and 2, all of this makes it harder to dissociate conceiving from execution while you are innovating—in foresight—than it is when you are looking backward on the path you followed—in hindsight.

In the examples in this chapter, "knowing it all" ahead of time is impossible, impractical, or simply pointless. What matters is being ready to face the reality ahead. You may find in your own experience situations in which you defaulted to a similar mind-set—knowingly or not. Think about situations in which you have found yourself working through changes to an idea powered by nothing but a feeling that something isn't quite right.

You can use that intuitive mind-set as a verification recipe of sorts. Indeed, it is the best metric to figure out whether you should continue working on something. If your problem does not compel you to adopt this mind-set, you can give up on the problem now. Since you don't really care about it, you might as well spare yourself some agony.

You need to make sure you introduce into that routine the same kind of existential anguish the question about getting killed supplies the astronauts. It isn't enough to get some confirmatory evidence or validate your ideas; you need to *verify* every aspect of your problem you can think of.

Validate or verify?

The English word "validate" comes from the Latin *validus*, meaning "strong," and is used to imply the gathering of evidence that makes your logic reasonable or cogent. "Verify" comes from the Latin *veritas*, meaning "truth."

You may have grown accustomed to using these words interchangeably, but they imply very different burdens of proof. A few confirmatory pieces of evidence can suffice to validate just about anything, even if it is knowably wrong. Conversely, you can never verify anything fully *a priori*. Say you want to verify the existence of customers. The proof lies in whether or not people buy. That lives at another scale, and when you get to it the question will no longer matter.

You verify by trying as hard as you can to prove the assertion wrong. When it persists in surviving, you begin to persuade yourself there is some truth to it. You should, nonetheless, keep on trying to prove whatever remains wrong as you progress through scale-up.

By contrast, to validate an assertion you need only gather evidence that suggests the statement is reasonable. Feel free to validate, but know that this might only lead you to finding out you are wrong later, when being wrong is more expensive.

You may have heard this phenomenon referred to as "confirmation bias." In truth, that bias is just a by-product of the experimental logic implied in validating. "Validate" and "verify" imply very different experimental logics. For the purpose of innovating, validate implies a useless (and, in the long term, expensive) set of mental operations. It is useless because no one cares about your "reasonable" statements; they want rock-solid ones.

If you go on simply validating, later rather than sooner things that are false will prove to have been reasonable. At that point, the best you can hope for is that a leprechaun will emerge from behind a rainbow with a pot of gold and show you the magical change in direction you need to take. It will be unplanned, unintentional, and serendipitous, but it will most certainly be welcome.

These kinds of pivots work best with leprechauns.

As Hadfield explains,

[Y]ou have to practice and learn what's the right thing to do. But given that, it actually gives you a really great comfort. It's counterintuitive, you know, to visualize disaster, but by visualizing disaster, that's what keeps us alive.

It *is indeed* counterintuitive. But it's not difficult to do; it is, in fact, far easier to do than it is to confirm that your approach to a problem is correct. The kind of mental operations Hadfield's description of the astronauts' process or my trip example illustrate can keep your problem alive and evolving.

Resolve that your "mission" will fail only in ways no one could have predicted. Approach it like the astronauts and "fail" at scale to avoid dying.

You can keep your problem alive and evolving by working backward from the problem being solved, posing questions, and trying to answer them by bringing parts and people together (either parts and people you "have" or you need to get)—in other words, prototyping your problem. Each time you do this, you will end up producing a kind of checklist of things that either have to be true or cannot possibly be true at the next scale. That checklist is, in fact, a plan. It is your best cue that there might actually *be something* to what you're doing—in other words, that you're on the right track.

What makes this endeavor somewhat counterintuitive is that instead of following a recipe to solve the problem like a robot, you have to trust that a solution will emerge from systematically negating the problem to come up with a better one. You need to engage in the kind of critical thinking that is a staple of heuristic techniques—that is, the process of gaining knowledge or some desired result by intelligent "guesswork" rather than by following some

> You evolve your problem—work backward, pose questions—and end up with a checklist of things that must be or can't possibly be true at the next scale.

pre-established formula. As I have learned in my teaching (and as I summarize in chapters 1 and 2), when it comes to innovating, not only is the alternative to heuristics—a form of robotic comparison ripe for a randomized control trial, such as the call for ideas I describe in chapter 6—impractical; it lacks power to overcome nonlinearity, and it mires you in hindsight thinking.

All you need to do is devise a *process of questions* that leads you into actively and tangibly fixing what you find is wrong with your problem. Talking to other humans or assembling parts, managing, and so on emerge as a consequences of trying to disprove that a problem that you assume solved is actually solved.

An inventory of the problem

In other chapters, I have had you assemble parts in a way you think you can emulate your problem. Here you're going to try *deliberately* to make the problem *not* work. Making your problem *not* work is, in fact, the only way to ensure that you cover the entire space within which your problem sits.

The questions outlined in other chapters lead you to an initial innovation prototype. That innovation prototype may have emerged from your first attempt at assembling parts and insights from people, it may be what's left from previous iterations, or it may have come out of an innovation prototyping kit you were given. Whatever the case, that innovation prototype represents a problem and should answer the following high-level questions that help make your problem more tangible:

- Can I reproduce the problem? Would I recognize a solution?

- What does impact look like? How would people encounter my solution? How would their lives change from using my solution? Can I prototype that impact?

- Would I be able to go over my prototype and explain the way in which people encounter it, the way in which it gets to them, and the benefit I claim they'll derive from it, or did I just prototype some gizmo? (This is a sort of "sanity check" question.)

- What is wrong about this problem?

- How can I fix it?

With this innovation prototype in hand, your best bet is to take for granted that this is indeed a scaled-down demonstration of a problem that can be solved and assume it will not work. That is, imagine all the ways in which it will not work so you can go on and figure out what will.

As you proceed through the next set of questions, you may discover that the problem you set out to solve is wrong in many more ways than you thought. As discussed in chapter 2, this may lead you to having to find a more accessible problem or an easier problem to solve first. The questions you pose to yourself and the answers you get from parts, scale, and people should allow you to do that.

You may end up with not just one but multiple, distinct versions of the problem you set out to solve. That's good; it would lead you to identify the kind of space of opportunity I discuss in chapter 5. Your objective isn't to solve the problem; it is to progress through an increasingly robust sequence of proofs of concept in an ever-widening and increasingly robust space of opportunity. As a matter of fact, interrogating your problem is a good way to outline what you aim to accomplish. I come back to this point at the end of this chapter.

A process of questions

In the example given above, the astronauts knew they were being successful if certainty came in the form of a new protocol or a request to change something. Similarly, you know you are being successful if your line of inquiry leads you to a new set of parts or a new set of people to talk to—that is, to a new set of "truths" about your innovating for which you need to create evidence. You may call these truths a set of hypotheses and experiments, but what's important is the certainty you seek to acquire about an aspect of your problem.

The questions you have to pose to yourself gain their power from the existence of a tangible demonstration of the problem. Absent a prototype, they may appear abstract. So, every time a question refers to a *this* or a *that*, you have to force yourself to imagine you are pointing to something in your innovation prototype. Conversely, if when posing one of these questions of your prototype you find nothing to point to, you know right then that your prototype is incomplete and/or that it is time to revisit an assumption.

Challenge what you think you've learned.
Broaden your proof of concept. You may end up
with multiple versions of your problem—a space
of opportunity.

You should ask questions about your knowns and unknowns that broaden your inquiry, explore proofs of concept, and challenge what you think you've learned. They should concern the problem as a whole. Their purpose is to help you begin a line of inquiry about what you actually have demonstrated so you can demonstrate something else—the next something. Questions ought to point to actions—that is, interacting with parts or interfacing with people—that you might take to answer them. The actions are what lead you to definite answers or, more precisely, to things you can check off. ("Do I have my passport? Check!")

What follows are some examples of questions that emerge from other chapters. This is not an exhaustive list. You will figure out many of your own questions based on the problem you're working on and what you learn along the way.

As I explain in chapter 2, you can count on being wrong with your current best guess about each of the things you do not yet know about your problem. You can, with near certainty, count on your imagined solution being wrong, and with high likelihood that your vision of what ought to be accomplished and how you'll verify a solution are also wrong. You can start your inquiry here:

- Can I imagine a way the problem may be solved that this verification rule would miss?

- Why is this solution I imagine not solving the problem?

Asking these questions of an early prototype may be difficult, so it may help to negate further what you have built by prompting your questions with any of the following:

- Why would anyone care about accomplishing that?

- Assuming the problem is solved, what is still missing if people are to accomplish what you foresaw? (This is a tricky question. The kind of answer you are looking for is the "such and such" in the following statement: "People will accomplish what you foresaw only if *such and such* is also true.")

- Something about this problem is not quite right. Can I play it out? That is, can I simulate the problem being solved?

These are all different approaches to a more fundamental pair of questions:

- Is this even a problem?
- If not, what is?

The second of those two may seem silly, but only until you ask it in this context.

These questions may get you started on a line of inquiry that reveals what you know and don't yet know about your problem given what you have prototyped. You need to sort what you don't know into what you can address now and what you'll have to assume to be one way or another for now and test later at a larger scale.

- Is that unknown important only because I imagine eventually solving a problem that affects lots of people?
- What remains unknown if I solve the problem for only one person?

As you progress, you ought to be able to recognize some of those unknowns as pointing to a need for new evidence—that is, data and conditions that you only now understand that you need.

- What do I need to be true for that thing to be also true at scale?
- What data can I obtain to show that it will not work?

Other unknowns are more difficult to formulate, so you may need to make them go away for now by modifying your problem, either by focusing on a piece of the problem or by taking your problem down to a scale at which the unknown is not an issue for now.

- What do I actually need to know now?
- What do I know how to solve, or can explain, with data, parts, or insights I have now? (Note that this ought to reveal the scale at which the unknown does not exist.)
- Can I make this smaller or larger in some way to overcome that unknown?

You need questions that will push you to consider everything you learn from parts or people as an auxiliary element (some of these questions appear in chapters 3 and 4). This allows you to focus your endeavor on developing partial proofs of concept—that is, on identifying combinations of parts and insights from people that together constitute proof of something, even if that something is only one of the many things you'll eventually need to be true.

You may begin to ask the following questions of every part and insight that has made it into your prototype. When interrogating your prototype at this level, it is useful to assume routinely that what you have is a near miss—that is, one small modification away from what it needs to be.

- What does this part or insight do for my prototype? Why is it needed? Can I remove it? What doesn't it yet do? What does this look like at a larger scale?

- What did I have to assume to use this part or insight? Is it still true?

You can then go on to explore combinations of parts and insights:

- What do parts A and B and or insights C and D represent when together? What do I need to know, add, or find out, so that they work together? What has to be true for them to work together? What other parts might I need?

You can achieve a lot by assuming that the outcome of putting things together isn't quite right even when it appears to be working. One way to reason your way into what else might be achievable is by posing questions that lead to new or changed parts and insights, which essentially amount to enhanced specifications:

- What part or insight do I have to remove or add to attain another specific outcome?

- Which part is not yet doing what I would need it to do at this scale?

- I might be able to trace a given outcome back. What is each part or insight really contributing?

Remember, in the context of this book a part is anything that is not a person. So, a business model (and its components) is a part, as is an imagined user (an archetype, a persona, or whatever). And you can operate on the insights you get from interfacing with people much as you do with parts. You should make parts and insights the subject of the same kind of inquiry you would any other aspect of your innovating, and assume they are but a place-holder for something else.

You need questions that facilitate coming up with the near misses with which to seize nonlinearities as opportunities (some these questions appear in chapter 5). So, as you go over your working innovation prototype you may make a point of modifying it by introducing radically distinct parts and insights.

- Looking at a specific component of your innovation prototype, ask these questions: What happens if I add a radically new part *here*? Remove *this* part? Modify the prototype to account for *this* insight?

A similarly unassuming line of inquiry may have led Edison to add a thin plate and then observe current flowing in one direction. The last set of questions seems to set you on a random search. It doesn't, though, because you are a highly evolved pattern-recognition machine. You'll find it difficult *not* to start seeing patterns that guide your search. Indeed, you are entrusting the search to your ability to recognize patterns. Even Edison's search was directed in that he focused on adding things inside the light bulb, first in lieu of the filament and then together with the filament.

You can follow a similar line of inquiry, looking at groups of parts and insights that seem to work together:

- What other parts or insights may this group work on as well?

- What if we assume *this* insight we took for granted is wrong? What should we change? What other thing might need to be true?

After you explore how to modify what you have built, you can ask about what you then have:

- What does this innovation prototype now *do*?

- What problem or problems is *this* a scaled-down version of? In what way(s)?

- Is this problem different from the one I started with? What specific change caused it to be different?

- What experiment or new set of parts or insights would I need to show that it will not work?

Whatever you are doing now, the result will be only a first semi-working prototype, and you'll need questions that help you synthesize what you have learned. That's a first step toward organizing the results of your exploration systematically so you can address increasingly larger versions of your problem—that is, toward systematizing your innovating, which I discuss in part III of the book. You can ask:

- What in my prototype suggests that this is an actual problem (solvable, recognizable, and verifiable)?

- What do I think I know about my problem? Can I inventory the insights and the large-scale reality captured by my prototype?

- How is it that my parts, working together, suggest the problem might be solved?

You know from earlier in this chapter that your main purpose is not to use these questions to find a solution to your initial problem, but to use them to identify ways to demonstrate that your first problem is wrong and thus find a better version.

The questions above should lead you to identify strategies you can use to modify your innovation prototype and make it represent different, larger realities. Your answers to those questions should point to a need for new parts or new insights with which to acquire new certainties about several aspects of your initial problem. The way you come up with one or several versions of your problem is to discard your initial formulation of the problem altogether and instead examine what these changes tell you about the things you now know how to do and the things you feel need to be demonstrated.

- What have I learned about the problem? Can I make the problem simpler?

- Are there patterns to the variations I applied to my innovation prototype?

- Can I group the experiments or next actions in some way? By parts? By aspects changed? By problem?

- What "questions" or "truths" continue to resist my attempts at proving them wrong? (Note that these may be the innovating equivalent to what scale-up engineers, control engineers, chemical engineers, and physicists call "invariants," "dimensionless numbers," "poles," etc.)

- Can I distill a new set of distinct problems from how I may group the next set of actions for my innovating?

Your objective is to define that space of opportunity (described in chapter 5) that may be covered by changing aspects of the problem you solve. In effect, what you are doing is searching through a space—exploring. You are successful in your exploration if the process of questions above leads you not to just one solution, but to understand the structure of the space of opportunity tied to that initial hunch. That understanding should translate into a family of diverse problems you can now explain in terms of near misses—that is, explain each distinct problem as a consequence of seemingly small changes to the set of parts that helped prototype some other problems. When one of those externalities that become apparent only at a larger scale presents itself, your understanding of the space around the problem that gives you purpose will make it easier to change aspects of what you are doing to circumvent the externality.

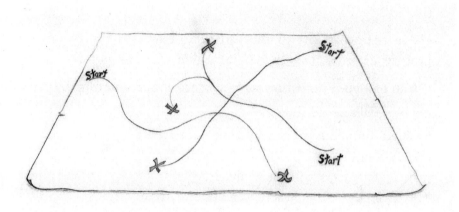

Your space of opportunity will have multiple opportunities, multiple ways to get to them, and multiple starting points. You have no way to know which opportunity is best without exploring. You chart your exploration so that the starting point does not constrain where you end up and you do not prematurely reject any opportunity.

Progress?

You know that you are progressing if at any point you realize that your problem is not worth pursuing, that the space of opportunity is smaller than you thought, or that someone else already solved the problem.

You also know you are progressing if your inquiry evolves somewhat gradually from a quest for parts into a quest for ways to make parts and insights work together and then into a quest for strategies to systematize what you have learned about your problem so you can serve a solution at a large scale. You also know you are progressing if after a while you realize your problem was not really a problem.

Progress comes from learning, realizing that your problem is not worth pursuing, or concluding that your problem was really not a problem.

Somewhere along that progression you'll realize that "execution" has overtaken "sensing." Exactly when that might have happened is hard to tell precisely in foresight; you can, however, arbitrarily pick any time in hindsight.

The set of questions I propose here is consistent with my presentation of innovating in other chapters. As you progress in your endeavor, you may find some questions more useful than others or come up with variants on these questions that are more adapted to the way you think. As you become skilled, you'll likely find that you need only ask yourself a few of the following questions to trigger your process:

- What needs to be true? What did I assume to be true?

- What if I added *this* or *that*? Or what if *this* or *that* was possible?

- Is this or that actually *known* to be impossible? Or is the impossibility just a by-product of how we think about it all?

- What do I need to put together—parts and insights—to show it all will not work?

Or you may come up with an alternate set of questions altogether that suits your own background.

General Attributes of Questions

The kinds of questions you need to ask yourself have a number of general attributes:

- They may appear to be very abstract if you don't have a tangible version of a problem, yet they are incredibly definite when posed about an innovation prototype. You have to force yourself to imagine that you are pointing to something in your innovation prototype.

- They ought to make it easy for you to imagine new truths—that is, things you do not yet know to be false. The point is to prove them wrong yourself through inquiry or through new combinations of parts. That's how questions can help you venture into an entire new space of opportunity.

- They ought to help you identify near misses.

- They ought to help you imagine variants to the problem.

- They ought to help you imagine scenarios under which your entire endeavor will be rejected. The point is to prepare yourself to translate the "you are wrong" answer you'll get as you converse with people and parts into a new set of tangible actions or combinations.

- They ought to introduce the same kind of existential anguish the question about getting killed supplies the astronauts.

Updating the problem and the innovation prototype through questions

As you interrogate your prototype, the answers you get from parts and people accrue into your prototype. Your innovation prototype should gradually reveal your increasingly sophisticated understanding of the problem. With your prototype, you ought to be able to explain tangibly the spectrum that spans from how parts should work together to how people encounter the problem and to how people stand to benefit from it being solved.

Your innovation prototype should gradually reveal your increasingly sophisticated understanding of the problem.

As you continue to find ways in which your idea is wrong, however, questions either will point to a need for new certainties or will reveal large unknowns that you may have to assume as externalities at this scale—though they may be addressable at a larger scale.

As your questions become more sophisticated, it may eventually become increasingly impractical to find ways to address them with the resources you have at this scale. As that happens, it should become increasingly easier to define what needs to be put together next (what parts and what kind of people) to pursue the "certainties" you can pursue only when you consider the problem at the next scale.

You can use all this information to update your innovation prototype. You begin by taking an inventory of what you have:

- a set of parts and insights you obtained from people working together and that illustrate something that may be possible at a larger scale

- a notion of how changes to those parts and those insights modulate the problem you want solve

- a number of questions posed and certainties that emerged from those questions about things that are wrong and about things that refuse to allow themselves to be proven wrong

- a number of questions posed and the unknowns they revealed, as well as the assumptions you had to make until later in the scale up when you would be able to revisit the questions

- new questions emerging from what you learned (things that may be possible given the new certainties you acquired) that you may answer with parts and people accessible at a larger scale

- several refinements of the problem that gives you purpose.

This is now your innovation prototype. You can summarize it in a different way by transforming everything you've learned into a set of logical statements that help you capture the possibilities you've encountered:

- If A is possible/true, so is B.

or

- For A to be true, B and C must meet condition X.

or

- After eliminating A, B, C ways in which X was to be proven wrong, we are left with D. Either D makes X wrong, or we need to find something that will.

and so on. These statements should all be based on the evidence you've gathered from parts and people or on the *explicit* assumptions you had to make. You have to work hard to make sure nothing is implicit.

With this you can try to restate your problem in a slightly different way: in terms of the space of opportunity it creates. You may assume your hunch really pointed to a space of opportunity and try teasing out several distinct problems—*new* problem opportunities—that span that space.

Each new problem should have the structure of a problem and should be supported by different aspects of your innovation prototype—as if composed of partial proofs of concept, parts, people, certainties, and unknowns. The new problems should be sufficiently distinct from one another so you can imagine them as resulting in altogether different solutions, but should overlap in that they all emerge from the same logical chain of possibility you outlined. When you progress to the next scale, every new certainty you

> Your questions should expand your initial hunch into a space of opportunity represented not by just one but by several distinct problems you can address.

acquire will translate into a certainty for one or more of those problems. As you continue to prove things wrong, you can think of recombining these problems into something different based on evidence.

Together, these new problems define and span the space of opportunity. Rather than pick one to focus on, you ought to determine the next set of experiments and actions that will give you the most information about the entire space of opportunity. You should choose the actions that will demonstrate soonest that the space is not worth pursuing as well as the actions that will give you the most information about the way in which the space might be interesting. Those actions constitute your logical chain of possibility. That is how you plan to explore and seize the space of opportunity.

The space of opportunity is the next refinement of your problem, and the sum of certainties and new experiments is the next evolution of your innovation prototype.

Takeaways

• Figuring out how to show that your problem is wrong is, in essence, how you work your way through innovating. As in the astronauts' preparation for a mission to the International Space Station, you ought to work through your innovation prototype trying to imagine the next thing that will "kill" your mission.

• The questions in this chapter allow you to get started innovating by assuming that you'll fail. As you become more skilled at innovating, you will make it a routine to assume that your prototype will not work, and to go systematically through all the parts and insights that make up your prototype—those you are using and those you are not using—to prove that it will indeed not work. That is, you search for a culprit that would turn an error at this scale into a failure at the next scale.

• You do that by interrogating your prototype with parts and with insights you get from people. Along the way, you may discover a few things that resist your attempts at proving them wrong. Throughout your search, you'll come to understand that these things transform your hunch into a space of opportunity.

• As you become accustomed to this mind-set, you ought to find it easier to add new parts and insights that you don't believe will work together and pose the same questions you would for the ones you suspected will work. A near miss is a near miss, whether by accident or by design. All near misses carry information.

• This is how you commandeer nonlinearities and make your innovating robust. Your objective is to build a space of opportunities represented as a set of problems you can explain in terms of modifications to your innovation prototype. The strategies you come up with to answer tangibly the questions your prototyping raises become your plan.

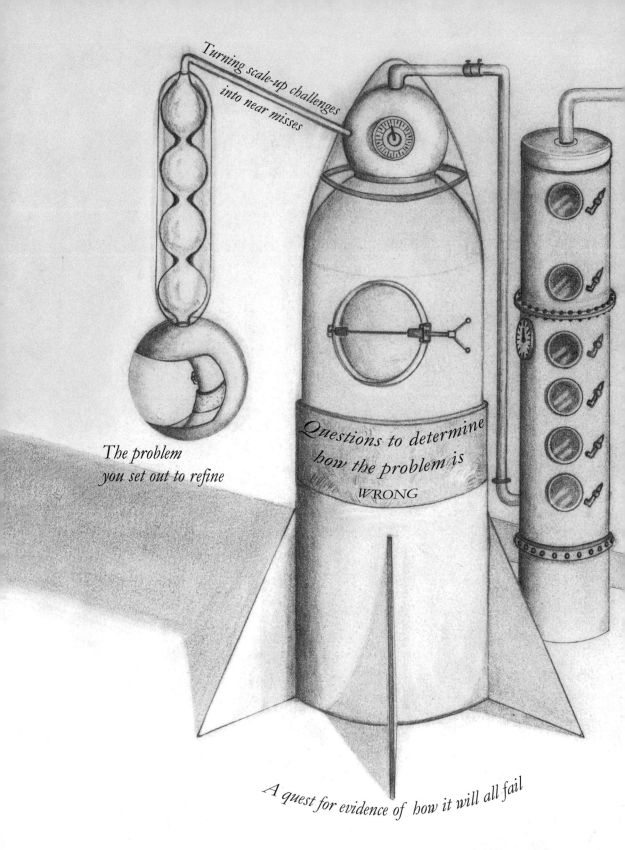

Turning scale-up challenges into near misses

The problem you set out to refine

Questions to determine how the problem is WRONG

A quest for evidence of how it will all fail

A new tangible set of questions

A refined problem,

perhaps multiple new problems,

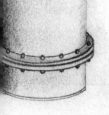

all creating a space of opportunity to explore

III

ORGANIZING WHAT
YOU HAVE LEARNED
Exploring Impact

IN PART II, I TOLD YOU ABOUT PARTS, people, exploring through nonlinearities, and "packaging" a hunch into a kit. I offered examples of questions you might ask along the path to help you find new ways to bring them all together to solve for a problem. The difficult thing to swallow might be that I have authorized you, indeed even asked you, to be productively wrong and not to limit yourself to just a product but rather to dare to explore seemingly preposterous ideas.

If you need some extra confirmation that there is nothing bad about being wrong and pursuing a seemingly preposterous idea, consider the three laws of prediction established by Sir Arthur C. Clarke (1917–2008), the great British science writer—both fiction and nonfiction—and futurist. The first two state:

1. When a distinguished but elderly scientist states that something is possible, he is almost certainly right. When he states that something is impossible, he is very probably wrong.

2. The only way of discovering the limits of the possible is to venture a little way past them into the impossible.[1]

However, there is a problem both in my explanation in part II and in these two laws themselves: They set you up in an infinite loop. Worse, although most innovation ideas begin as preposterous, not all preposterous ideas graduate into becoming real. (There are numerous books on the subject of patents that never became anything tangible.) There is no guarantee that your idea will graduate.

It gets worse. No method tells you how to figure out when it's time to stop "iterating," whether your idea will graduate from being preposterous but possible to becoming probable and doable, or even whether you'll be the one to graduate your idea.

Clarke's third law gets you out of the infinite loop by postulating that at some point the elderly scientist from the first law will concede that you've conjured your way around the "impossible":

3. Any sufficiently advanced technology is indistinguishable from magic.

Back to innovating: You begin and end with a problem. There's your infinite loop. If you keep going, don't expect there to be a big, bright, red stoplight. Nothing you do will reveal an exact stopping point. To be sure, you are not a robot or a slave. Stopping is a decision you make, and you can stop at any time. You might pass your innovating on to the next person as, say, an innovation prototyping kit. If you do keep going, your way out of the infinite loop is to grow.

As you learn, you hope to find ways to simplify what you've already done so that doing it again (should that be necessary) takes less effort. That's how you grow. Simplifying frees you to approach the problem that gives you purpose at its rightful scale.

Put another way, you set out to organize the outcomes of your exploration systematically so your exploration takes less effort (measured in money, time, and or skills) to do it again. That is, as you set out to systematize continually what you learn, some organizing principles emerge that form the basis of an organization. As that organization emerges, you finally stand to benefit in a strong way from what the field of management has laid out for you. I wrote about these highly specialized concepts in the introduction to this book:

Keywords and highly specialized concepts such as need, product, distribution, value chain, users, lead users, competitive forces, value creation, and value capture do not have meanings set in stone. At the beginning of an innovator's inquiry, they are largely undefined and ambiguous; they acquire their precise meanings and their analytical strength only over time through the inquiry of the innovator, from the organization that emerges, and in the context of the problem that organization ultimately solves.

This should clarify my reference to an infinite loop. You set out to solve a problem that ultimately requires some kind of organization. The organization ought to be solving the problem continuously for as long as the problem exists—that is, forever or until a challenger ousts you. That's what the innovation literature predicts. That might happen, for instance, when the problem that gives you purpose has become obsolete, that is, you stopped solving for the problem—you stopped innovating.

A better question than whether to stop is what you want to grow. That begins with systematizing what you have learned, which you can choose to start doing at any time. You do not have to wait for a revelation to emerge from anything you do.

Here are four examples of things you might want to grow that could go on forever:

- You could make a life's work out of turning hunches into increasingly clear opportunities. This might involve, for instance, building an "innovating corporation" that specializes in identifying tangible opportunities and "spinning out" organizations, much as research labs "spin out" papers.
- You could increase the scale of your innovating incrementally and make external investment a choice. People have been known to build large businesses this way; chapter 7 provides the logic that would allow you to do so.
- You could make it a point to build one business and then build another on top of it. In chapter 9 I show you how to advocate for

resources to do this, and in chapter 10 I show you how to think of growth this way.

• You can also set an arbitrary time limit for resolving whether to jump a scale up or move on to a different hunch or different people or different parts.

The chapters in part III help you look at your innovation prototype for the purpose of systematizing what you learn. Chapter 8 helps you understand advocacy as something you've been practicing all along. Chapter 9 explains how to view risks and certainties as progress indicators for what you have learned and as instruments for communicating the robustness of your idea. Chapter 10 explores the notion of growth in connection with scale. Chapter 11 shows you how to enact this book as a process with teams engaged in innovating continuously. Chapter 12 explains how to document what you do so you can get started again at the next scale or from the very beginning—it makes no difference.

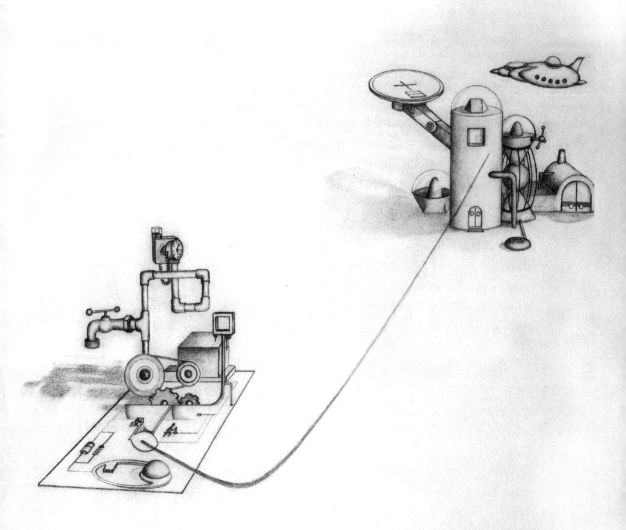

PRACTICING ADVOCACY

At some point, it may seem as though everything your problem calls for next requires resources you don't have: parts beyond your means, skills you lack, money you don't have, and/or knowledge and expertise you don't possess. It may feel as if your problem is conspiring against you to erect a roadblock.

Put another way, it may seem that taking the scale of your innovating up a notch is the only path left on which to make progress. You don't really know whether that's true. But there's only one way to find out: Approach people who have the resources you need and propose that they join you; and continue to make it a purpose to find out how you are wrong. In other words, keep doing what you should have been doing all along.

The juxtaposition of two ideas—you do not know how to progress and yet you have resolved to advocate for resources with which to make it even bigger—may feel like anxiety, and more so the first time around. You know it is anxiety because unanswerable questions begin to race through your head. Should I seek funding? What if there aren't any investors where I live? What if this is my only shot with the senior executives? Do I just need to put in a bit more work into it?

The way out is to assume you probably have all the connections you need, the people you need *do* exist and *are* accessible, what you have ought to be enough, and—since these questions have emerged—now is probably a good time to go out and seek those resources. There is just one assumption that is wrong: you are *not* going out for resources to make your idea bigger.

You never get only resources; people come along, too. Together, you build a *bigger idea*, not just your bigger version of your earlier idea. You are building a deal.

You never just get resources; people come along, too. Together, you build a bigger idea, not just your bigger version of your earlier idea.

You now see a way to solve a problem that is impactful and interesting at a larger scale, and you have a good sense of what it will take because you can itemize what you are now missing; you also know that problems evolve as they are influenced by the people who work on them.

Let's look more closely at how all this works. Whether you are advocating, verifying, or inquiring, you focus on different aspects of the problem you have, and you ask for different things. Advocacy probably requires the most comprehensive look at everything you do, so let's begin there.

You may take advocacy as an opportunity to practice synthesizing your innovating as a whole, and reveal your innovation prototype as what it is bound to become to others interested in knowing how you propose to use their resources. It is a great exercise, and you should make a habit of it: Restate your problem by making your innovation prototype be a "demo" of the problem solved. Restating the problem invariably reveals shortcomings and holes in your reasoning that you can generally remedy through inquiry or verification. That is, preparing for advocacy is valuable in itself.

You can begin by thinking that in all these activities—advocacy, inquiry, and verification—you likely know more than your counterparts about the idea that gives you purpose. They have resources you need, but they may not even know they do if they can't understand you. Your job is to make the problem you've made tangible for yourself equally tangible for them (within time constraints) so you can focus the discussion on the things they have or know and you do not. That's when all the work you've put into your innovation prototype as you strived to make the problem tangible to yourself becomes extremely valuable. The more "tangible" the conversation at this point, the greater the value.

Your purpose is not to seek approval or get permission, even though that's what you might get. Your purpose remains—as always—to find out how you are wrong. However, if you are seeking funding at this point, everything may feel a bit different.

At times, entrepreneurial recipes place so much emphasis on funding your idea that it is difficult to distinguish advocacy from *propaganda* when it comes time to communicate with others who might be sources of that funding. What you need to do instead is communicate with your counterparts confidently, truthfully, and in the spirit of learning how you might be wrong.

People may take you for someone just asking for money if they can't understand you. Your job: Make the problem as tangible for them as it is for you, so you can focus on things they have or know that you don't.

> Your purpose is not to seek approval or get permission. Your purpose remains, as always, to find out how you're wrong—even when seeking funding.

Then you build on that communication to establish the basis for trust and for getting to know one another. And then you progress—if appropriate—toward making a deal.

The deal you are about to strike goes like this: You are inviting some people with resources into your innovating in exchange for some of those resources; they likely will want something too. Before any of that can come to pass, you need to get their attention, and you all need to get to know one another.

Advocacy as I am describing it here is that communication about your idea and a space of opportunity, that getting to know one another, and getting to a deal.

You've Been Advocating All Along, and You'll Continue

Advocacy is tricky. Early on, you may feel the pressing burden to fund your innovating, turning the sole purpose of advocacy into coming up with a way to get investors of *any* kind to love your idea. However, advocacy is something you'll have to do several times, for several different

---

purposes. You'll have to advocate when you are looking for customers, when you form a partnership or a joint venture or collaboration, when you go through the next stage of growth, and even if (or when) you decide to sell your business.

In fact, you've been practicing a lot of what goes into advocacy all along. Each time you've shared with others the problem you are solving, or some aspect of the problem, you have been practicing skills that play a role in advocacy. You've been practicing with every inquiry about your problem, such as seeking some specific knowledge you need or capturing unanticipated insights from people's comments. You have also been practicing when you engage others to *verify* aspects of the problem to get a sense for whether your efforts to solve the problem match reality.

You can view advocacy as the activity you have to systematize and perfect as your endeavor evolves from a subsidized business (e.g., a startup) into a self-sustaining one. This is also true if your endeavor is social: Eventually, your advocacy for resources will have to evolve into an organizational unit that makes sure you can continue to operate at each appropriate scale.

You may be advocating to an investor, strategic partner, owner of an exhibit hall, large corporation, acquirer-to-be, or donor—to name a few. In all cases, what you propose is the same: Given what you have and what they have, there is a way to build something new that makes everyone better off.

Those to whom you advocate don't yet know what makes what you have interesting. You don't really know what they have, nor do you know what they need to know. Hardly anyone decides to hand over resources in one sitting.

So, prepare to build relationships—to *advocate*—through several conversations, as you've been doing all along.

In this chapter, I discuss how to view advocacy as an opportunity to synthesize the problem that gives you purpose so you can communicate it. I see advocacy as a natural extension of the skill you've been acquiring all along as you've practiced interfacing with people.

Advocacy and fundraising

You already have the basic skills you need to go out and advocate for your innovating to obtain the resources you need. Again, what might feel different this time is if you've persuaded yourself you need funding to continue your endeavor. To many, the perception that funding is needed can trump everything they have learned and demonstrated, propelling them into fantasyland enabled by slide deck. Aspiring innovators can follow myriad recipes for how to pitch an idea, talk to potential users and customers, and engage in similar activities. When the time comes to advocate for their projects, however, those who follow those recipes end up forgetting that the objective is, first and foremost, not funding but progress.

This happens for two reasons. First, most entrepreneurial recipes inadvertently have you adopt a rather idealized duality: You have the idea; they have the money (or geeks and suits, or the innovator and the user). These recipes supply you with useful ways to craft a story to please your audience and persuade them to part with the funds you covet. In other words, they help you beautify and market an idea, which is different than—though often confused with—bringing your idea to market. In the former, you sell "the idea"; in the latter, you sell something that was at some earlier point *only* an idea. But those recipes don't prepare you for the reality that those people do not just give over their resources, but also generally trade them for some kind of ownership. It could be equity ownership or ownership over collaboration—in either case, you *both* own the idea at the other end of the deal. Second, those recipes take for granted that you've already done whatever you had to do to make your idea good and solid, and that you are confident about it. All that remains is execution, which by now you're also supposed to have figured out. Here again, hindsight factors in: The distinction between when you stopped sensing and when you started executing is something you can *really* identify only in hindsight. I know of no one who has that clarity when engaged in advocacy for innovation. But because these recipes focus overwhelmingly on presentation, it seems at times that they are telling you to put *doing* on hold until you've gotten the resources. Even worse, these recipes suggest that *doing* prior to advocating to investors, users, or whoever somehow disqualifies you.

The Fine Art of Letting a Recipe Get You Lost

At this point, most entrepreneurial recipes that have asked you to bet it all on a product ask you to give it a name, make a statement about the problem you are solving, and build your value proposition. I don't necessarily disagree with that general flow; the recipes might work. But describing a problem is difficult enough without doing so merely for the purpose of making your product appear unique.

So, what should be your focus? A problem or a product? If you focus on a problem, the product must be subject to change. Most people evade this question by supplanting the problem with a dramatic or even a catastrophic fact (e.g., "783 million people do not have access to clean water") and hoping that it passes for a problem.

Building a value proposition is also difficult. What you are proposing is obviously a proposition, but people to whom you advocate—whether it be to sell a product or to raise funding—expect you to have done your homework and to present something that is valuable to them, *valuable* being the operative keyword.

The same goes for the name of your product. At first people just need a handle, something to call you. A brand can be a handle, but a handle need not be a brand. For instance, you could just assign a random name— say, Project 71. For a while, it simply will not matter.

If you've never advocated, or your idea is just a bit audacious, it is difficult not to get yourself and your innovating lost in a "pitching" recipe.

The idealized duality view doesn't help you much, because it sets you up to believe from the get-go that you need them more than they need you. After all, they have money and you don't. Even if that's true, it's a poor negotiating position and one you should avoid. Chances are they have at least some interest in putting their resources to work on innovations; otherwise you would not be meeting. You have demonstrated something to be possible and probable, and you have a plan to make it doable and sustainable. You need to believe you'll pull through no matter what, which includes believing that you'll be able to do it with or without them. If it isn't them, as I already noted, you'll find better associates down the road.

Still, when you advocate—even for funding—you should view your counterparts as people, not wallets.

I suggest you take a skeptical view of these idealized recipes; that's how you'll get the most out of them. I see advocacy—along with the related activities inquiry and verification—as part of a continuum. All these activities are engagements with other people. Each of them affords you an opportunity to practice important skills. All of them help you develop the kind of "muscle memory" that comes from practice.

If you've never advocated, or your idea is just a bit audacious, it's difficult not to get yourself and your innovating lost in a "pitching" recipe.

"Automating" Your Advocacy Through Practice

In this chapter, I emphasize more than in the preceding chapters how you go about several aspects of your innovating in a manner that helps you become better at doing it. Earlier, I described it as a kind of muscle memory.

Doing things in a seemingly "automatic" way—that is, without consciously giving it some thought—is something we adults describe with a lot of different words. I think this comes from a need to rationalize to ourselves those moments when we operate "off script" and without a formula. We call it "intuition," "creativity," "bright idea," "connecting the dots," "insight," and so on. We say it can't be explained, but that we know it when we see it. Most of the words we use to describe this phenomenon have positive connotations, implying that we generally like the outcomes, but they are also somewhat abstract and hindsight-heavy, which is why we shy away from using them when called upon to explain to someone else what to do or how to do something.

The basic idea behind making something you do "automatic," though, is quite simple: The more you repeat a task, the more your brain seems able to "automate" aspects of it so that each time you undertake the task you need to expend fewer resources. That frees resources for learning the next level of skill. Also, the more you can find similarities with some other task you've done before, the less you'll need to "compute" how to do the new task and the more you can rely on what you've already learned.

Research suggests the human brain indeed works this way. Does it have to do with our ability to acquire language? With internal grammar? With conditioning? The good news is that you don't need a scientific explanation for how or why to use this to your advantage.

Finding similarities in tasks helps you practice so you can benefit most from this way your brain works, without relying on formulas. This is how you equip your brain with whatever it needs to produce intuitions later. Over time, in school, we generally forget to do this, because we're taught so many specific formulas for this and for that. But your innovating will be well served by finding similarities among different things rather than emphasizing differences.

Advocacy as an extension to inquiry and verification

At a high level, advocacy, inquiry, and verification all support the same objective: A very specific "something" is somehow preventing your path to scale-up, and you need to overcome it in order to continue. You do not have the means to do so, but "someone" else does. You discover what the "something" is that you need as you make your problem tangible; you need to engage and persuade the "someone" who has it to give that "something" to you and see what he or she may want in return. The "something" might be information, insight, contact, a part, knowledge, or money. The "someone" can be any of the people you encounter in your innovating.

That engagement involves the following:

- You *talk* about the problem you've made tangible, and why it matters—which is the answer to the "what are you up to?" question from chapter 4.

- You *state* what is needed to progress. Depending on how you frame the story, this may come from how you verify the problem is solved or from what any solution needs to accomplish.

- You *show* a tangible demonstration of your problem—that is, your functioning innovation prototype that illustrates the larger reality you aspire to conquer.

- You *ask* for the resources you need to take your next steps—that is, the next set of parts and insights that emerge from your interrogation of your prototype.

- You *demonstrate* as tangibly and sensibly as you can how what you've asked for will translate into the progress you said was needed—that is, you explain your next steps, the logical chain of possibility discussed at the end of chapter 7 (on which I expand in chapters 9 and 10).

- All along, you *listen* in order to find out how you are wrong. By now this should be a habit. You don't accept comments at face value; rather, you argue with facts and drive the conversation to what in the opinion of those you converse with would make it all right.

Let me show you how this may work and how it connects to inquiry and verification by playing out the case in which you are advocating for funding and trying to figure out what might go wrong—as in the trip examples in chapter 7.

Getting ready for advocacy

You've decided to add a task in your innovating: to seek someone else's resources. You will have to explain to others what you have, where you are headed, and how their resources would come into play. You either show them what they might accomplish by joining forces with you or you leave them confident that you will be a good steward of their resources—that is, you intend to use them as you use yours and learn a disproportionate amount from doing so. You will also need to discover what they might want from you.

That's a lot for a single conversation, so plan for it to unfold over several meetings. Your objective is to learn as much as you can each time. But first you have to get their attention. As a matter of fact, you'll have to continue to earn their attention all along until they care enough about what you'll do together that they see your innovation prototype—as you do—for the larger reality it illustrates. That larger reality likely will be slightly different from what you see now. Put another way, what you have now is a near miss of what you'll have after you all agree to work together. Right now, though, they don't care.

Let me emphasize that. They do not know you, they do not care about your innovation prototype, and they do not care how you got there—until you earn their attention. You need a story.

You do know the *kind* of story they want to hear: You are proposing something that will create significantly more value for both of you than what a bank might yield for an equivalent investment—and then some to cope with risk. This is obviously true for the case of professional investors, but it is also true for partners, or even suppliers and customers: when it comes to innovation, you buy into things—and also buy things—that you believe will make you far better off in some way. Even at home, unless you are paying the utility bill, you hope that the things you buy will give you back more than they cost you. That's a quick way to define value.

You can sum it all up in the first advocacy prototype. The curve shown in figure 8.1 illustrates the dynamic of your story, the magnitude of your impact over time. Your particular story determines what goes into the axes, the units, and the scale. Adding them is your job.

For instance, you can gauge the magnitude of impact by sales, number of views, dollars, number of citations, or number of visitors. You might gauge progress toward scale by time, milestones, or some other measure.

Since it is up to you to define all the axes and scales, you still get to determine how far the right end of curve will go. Note, however, that the less you

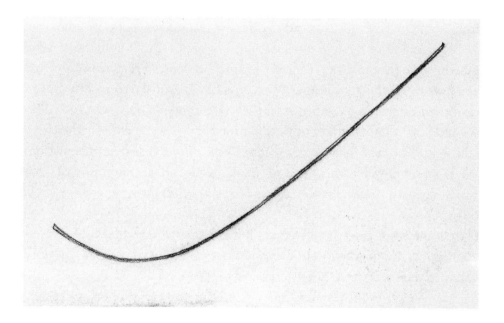

Your growth curve, for which you define the axes.

aim to produce at the left tail end, the less you implicitly value the effort you are putting in. Do not expect anyone to value that effort any more than you do.

Just don't forget that the point of doing this is to begin appraising the magnitude of the job you are giving to yourself and your team for the next several months. It is *not* to paint a rosy picture for prospective investors. You are planning backward from the problem solved. The team is committing to bring the solution you envision to the scale at which it really is a solution. You are all now innovation scale-up engineers.

Whether you choose to show the curve explicitly is a matter of style. However, it helps keep you honest. Whatever your solution, and whatever your problem, your story arc needs to make sense. You have something and you are asking for something else; it will take some expending and some time before what you do yields any return; the return over time will be significant; and your story needs no strange or hidden plot twists or ridiculous leaps of faith to get to your conclusion, which is a demonstrable impact.

This is also a story about learning and organizing. It takes more than a handful of people to build and grow something. If you are to learn a disproportionate amount and grow, you need to systematize everything you learn so it benefits the people who will join you. Your learning will translate into some specifications and organizing principles, and what you are growing is an organization.

Whatever your solution, and whatever your problem, your story arc needs to make sense. You need no strange or hidden plot twists or ridiculous leaps of faith to get to your conclusion, which is a demonstrable impact.

A Business (Organization) Model

The organization you are building is a machine. It creates order from chaos. You need to put work into your organization or else it violates the second law of thermodynamics. For our purposes, it suffices to know that the law's main point is that nothing gets organized by itself.

No matter how intricate that machine, you ought to be able to explain with a simple model how it will organize things. The model must explain the dynamic flows of people, capital, and resources from the current state to a problem being solved.

Since you are building the machine while running it, your model will also have to consider how the machine evolves. This can get hairy for a model with only arrows; you may need to run multiple scenarios in a spreadsheet for things to begin to make sense.

That's the model of your *organization*. If you built a business, it is a business model. And at the very least it ought to explain how you'll pay for things, how things will get where they need to go, how beneficiaries will find you, and who will pay you and how.

There are various recipes you might follow to analyze the structure of a business by its components, organizational finances, and/or tabulated comparables (with names such as "razor," "crowd," "subscription," "advertising," and "reimbursement"). But you need to have conceived of the dynamics of your organization before they're of any use. A model is fundamentally an explanation of how those dynamics work.

In my teaching, I have found it useful to build toward an understanding of organizational models by addressing first a business model's purpose: You are proposing to arrive at a state in which parts and people are increasingly organized and a problem is solved for a community of people outside your organization. This increase in order is also a decrease in entropy.

James Clerk Maxwell and other physicists explained that even demons cannot escape entropy.

This logic applies if you're building a business (which has been the focus of this chapter so far), if you're an artist, if you're a researcher, or if you envision a nonprofit. An artist, for instance, might build an organization through exhibits and might measure growth by number of visitors. That artist can only hope it will not take as much effort per person to get the first ten visitors as it will take to tend to millions.

To summarize getting ready for advocacy, this is what you're trying to accomplish:

- You are growing an organization that will solve a problem sustainably.

- You have an innovation prototype that demonstrates the problem and, by demonstrating, outlines the organization.

- You are asking people to join you in your innovating and to bring their resources (money, insight, direction, contacts, experience, ...).

- You have a proposal for how to use those resources and what they will achieve for all of you, and you see a path to a disproportionate amount of learning (a.k.a. value) using those resources.

Advocacy meetings: a ballet in three movements

After a number of meetings—in my experience, it takes at least three and probably more—you'll either have a deal or not. From meeting to meeting, the "resolution" at which you discuss your innovating will become sharper and sharper. If you make a deal, it likely will modify the axes, units, and scale of what you initially proposed.

The meetings are like a ballet, and ballets require choreography. This may seem like an odd way to describe meetings—it certainly is nothing like the entrepreneurship recipe books—but taking this approach, I believe, shows how accessible the process is and avoids the tedium that is inevitable when describing meetings in more traditional ways.

Allow me then to "choreograph" what meetings may look like by analogy to a three-movement ballet. Admittedly, this is a fantastical ballet in a very extreme setting. You are unlikely to set out to talk to someone who is a complete stranger, as in this ballet. It's much more likely that you'll "waylay" someone who knows someone you know and you'll drop the other person's name. And you probably will do it all in a different setting. But please indulge me this story—inspired by the film *The Hudsucker Proxy*—and give it a chance to brew in your mind.

In the ballet, you are the prima ballerina, equivalent to the character played by Tim Robbins in the movie.

Movement 1—getting someone's attention

The curtain rises on the first movement of our ballet, in which you find the person with whom you want to have a meeting. Let's say he's an executive at a company with headquarters in an extraordinarily tall building. You encounter him as he gets on the elevator and pitch your plan for world domination during a very slow ride to the top floor. He exits the elevator and bids you farewell, shaking your hand. You notice, however, that his eyes and body are already aiming elsewhere. Still, you hold firmly to his hand and continue to talk about your idea at an increasingly faster rate until a security guard notices and intervenes to undo your grip. As the two men rush away, you shout out your request for another meeting, go back down on the elevator, and continue to ride it up and down as you await the moment to pounce on your next prey. The curtain closes on movement 1.

In some other ballet, you may have succeeded in getting that meeting. In this one, you did not.

What might have gone wrong? It could be that you forgot to ask for advice, or perhaps you failed to mention that there is a purpose to your reaching out—say, getting funding—because you were so focused on getting out your idea. Maybe you came across as too nervous, which made you sound doubtful about your own idea. It could be that you talked too much about what you had already done and not enough about the impact or about how you envision a future to which your counterpart might contribute. Did you fail to make the problem sound interesting? Might the problem need something else to interest others?

Alternatively, you might have lucked out: He told you why he sees your problem as uninteresting. That will help you be better prepared to describe your idea. Describing an idea that is new and only half-formed is quite difficult. You can make it easier in two ways. One begins with accepting that, even though your prototype has helped you imagine some larger reality, it is only a demonstration of what's possible. The other is by practicing your explanation until you feel it is second nature to you and you can present it clearly and confidently. A lot depends on how confident you sound—much more so than on how robust your idea is.

The last point is important. Prototyping something gets you deep into a problem, and it can be tough to come back up to the surface to make an easy pitch. That's part of why the recipe books prescribe that you not *do*. But with some practice, you can easily accommodate in your head both the deep view afforded by your innovation prototype and the confident, synthesized description of a larger reality in which the problem that gives you purpose has been solved (even if that latter description is almost surely a near miss).

If you were even luckier and the executive asked or said something about your idea, he cued you to ask for a more formal meeting. Even if he said he wasn't interested in funding you, you might succeed in persuading him to meet with you to give you some advice or to point you to someone else—and thus transform what started as advocacy into an opportunity for inquiry.

Practicalities for Getting People's Attention

Every one of the encounters described in chapter 4 is an opportunity to practice. You should make a point to create different versions of an answer for the "What are you up to?" question and try them out. Now the prototype is communication, and words are parts. You know you are ready to explain what you are after when doing so becomes second nature—so much so that you don't even think twice about it—and when, in every single case, people ask you follow-up questions about what you *do*, not about what you *meant*.

If you are indeed selling an idea, then there is a value to the "sale." It just so happens that what you are selling now—a piece of your imagined company—may itself do something that also has a value proposition. However, for the sake of getting people's attention now, it is the first "sale" that matters, not the value proposition of what your imagined company sells. The value proposition of what your company sells is given by what people will be able to accomplish when the problem is solved. The value proposition of your idea is the clarity with which you can articulate what needs to be true about a solution—your verification recipe.

My research shows that most people, when recounting to others a presentation they just heard, find it sufficient to use only three ideas. If it's your presentation they're recounting, you want them to spread the word about how valuable a solution you proposed—even if you didn't describe in detail how you intend to solve the problem.

You get people's attention with a teaser. The feature presentation is in movement 2 of the imaginary ballet. You outline the value proposition for joining you in the adventure that is to follow, which hinges on the problem you solve, what you have demonstrated, what you are asking for, and the magnitude of the impact you foresee.

Making Your Point

Make your point first. If you are asking for something, ask. Be sure that your materials are always structured so that it takes no more than a few minutes to give a high-level description of your purpose and what you are after. Prepare, rehearse, and practice so that your description will come out natural and clear. Then be ready to expand on what you described as the conversation unfolds. Allow yourself to be interrupted; in fact, hope to be interrupted. Give conversation a chance.

Let me tell you how I came to learn the importance of all this. When I was a PhD candidate, I attended a training seminar titled "Introduction to teaching for non-American teaching assistants." The instructor drew two triangles on the board, one pointing up and the other down, and said that time started at the top of each of them. Students, she explained, expect you to make a point early on, and then to elaborate. That gives students the opportunity to figure out what you're talking about and gives them license to interrupt with questions if something isn't clear.

Referring to the other triangle, the instructor explained that many non-Americans try to build arguments until the point becomes apparent. Do that, she explained, and the students have to wait until you're finished before they know what you are talking about. Sure, they try piecing things together while you're speaking, but that only shifts their attention away from whatever you say next, and once you're finished they have to try reconstructing all the questions that emerged. Teaching this way leaves any classroom interaction dependent on students' ability on that day to hold onto your every word. In view of how little students sleep, you may never get your point across.

When teaching, she concluded, use the other triangle. Make an effort to make your point first.

Movement 2—the larger reality you will create together

In the second movement of our ballet, you've succeeded in getting a second meeting. The executive you waylaid in the elevator meets you at reception and takes you to a conference room, where he introduces the company's chief innovation officer. She tells you she's intrigued by the plan for world domination, which she recounts in three strokes.

You make your presentation, in which you demonstrate your prototype and show a bunch of charts—some intended as inspirational and some with carefully constructed projections of what's going to happen. Most of what you present are educated guesses. There are a lot of questions. The curtain closes on movement 2.

Every question asked matters. Every question points to something people expect you to know, comparables people use, and essentially everything you can derive from your inquiry. Unlike in earlier inquiries, though, here you've shared your entire prototype, and so feedback is likely more targeted than what you got from casual conversations or what you got from asking an "expert" about a specific aspect of your problem.

Comparables?

At some point in movement 2 of the imaginary ballet, comparables may come up. Their purpose is to give tangibility to your larger reality. They illustrate how others (such as other companies) in your space do things and, by analogy, how you might have to do things.

As you prepare for advocacy, there's value you can extract from any kind of comparable. Comparables can be parts. Everything in your innovation prototype that is not a material part may be explainable as "we will do such and such, not unlike Company X." To be clear, Company X may not be an actual competitor and may not even be in your space. Used as a part in this way, a comparable helps you render an aspect of your problem tangible.

Here's how you can use comparables as parts:

· Pick any company.

· Ask yourself what you think it does that resembles what you want to accomplish.

· Figure out how the company does it.

· Try to show you are wrong, that is, that you can't replicate what it does.

You can also identify these comparables as potential acquirers, collaborators, and partners in this way. After all, you claim you want to do something like they do. Maybe it is more effective for you to do it with them.

But even if a comparable ends up fulfilling no further role in your innovating, it will still make your idea more tangible and help you systematize. Whether you include the comparables in your story explicitly or simply talk about what you will do on the basis of what you learned about what they do is a function of how you choose to tell your story.

Maybe the meeting was good; maybe not. Let's assume it was a catastrophe. What might have gone wrong? Perhaps your feature presentation in movement 2 felt unlike the teaser from movement 1; maybe you had to resort to unexpected plot twists to make the logic work—a leprechaun. Perhaps your "comparables" made no sense to your audience. Perhaps, as often happens, what you demonstrated was not even needed in your story.

Perhaps no one agreed that what you proposed solves the problem you described. This also happens frequently—after all, you can reduce the most grandiose problem to a slide deck and propose next steps. You'll know you've fallen into this trap if your audience fails to see how those next steps lead to the problem being solved.

Chances are your narrative follows from what you have prototyped going forward rather than backward from what you envision building. Your innovation prototype is but a scaled-down version of a problem and its solution, and your job is to scale it up. Your story needs to project backward from the future you plan to build so your innovation prototype is a proof of concept of what's possible—no matter how you arrived at the prototype. It needs to do that because you should not expect your audience to do your homework for you. Ask yourself the obvious question first: If neither you nor your prototype/tech/bizmodel/design are in the picture, might the problem be solved more simply? Perhaps with a slingshot?

Earlier in this chapter, I wrote the more "tangible" the conversation, the greater the value. It applies here: The more tangible the reality in which you anchor your story, the higher the perceived value of what you would contribute to those who hear your story, and the clearer the value of what they might contribute going forward.

Mind the trap: Grandiose problems fit in a slide-deck, next to steps you'll never walk. Use your innovation prototype. It demonstrates a tangible destination.

Your innovation prototype can be that tangible anchor. It demonstrates a final destination that does not yet exist, and helps you illustrate a foreseeable impact. You can show the functions of your eventual organization that you deemed critical to demonstrate at this scale, and you can point to tangible next steps and milestones, the associated risks and certainties, and the resources they require. I discuss these in greater detail in chapters 9 and 10.

How Your Prototype Helps You Tell Your Story

You now have an innovation prototype that makes an entire space of opportunity tangible. Use it to help tell your story:

• Convince your audience the problem is real. Your prototype demonstrates this by means of what needs to be accomplished, what makes a solution a solution, or through an outline of a solution. The information you gathered hints at the impact of the problem being solved and the specific need to attain a certain scale.

• Show your audience the problem can be solved. Your prototype is a representation of the organization and the people required to solve the problem at scale. You can use it to outline precisely what is needed to address the problem at scale.

• Demonstrate to your audience how to reach an endpoint.

Because your innovation prototype also addresses the organizational parts needed to address the problem, you may use it to make apparent the vehicle with which you solve the problem at its true scale. You can show how you imagine the endpoint and then work backward to show what must be accomplished.

The specific objective of your advocacy is to enable you to continue scaling up. It is not to get permission. Even with a tangible anchor, there is no guarantee others will give you what you need. They may not even have it. Perhaps they do, and even consider your logic sound, but still choose not to give it to you.

There are two possible outcomes at the end of the ballet's second movement: You will or you will not get the specific "something" you were after. If you don't get it, the executives probably said something that reveals why. That valuable insight may not apply immediately, but it will apply eventually. Either way, you walk away from that conversation empty-handed only if you believe you have everything figured out and that the others are the ones who are wrong.

Movement 3—the specific details for a deal

Having explained your plan for world domination and further aroused the interest of the chief innovation officer, you have succeeded in getting another meeting with her and the other executive. (In someone else's ballet, there might also be lawyers, accountants, and others at the table.) Now it's all about specifics. They ask a lot more questions, but the nature of the questions has changed. The questions are no longer about your prototype or the larger reality you envision at the end; now the questions are about how to get there. You recognize this as "due diligence."

After a while, the chief information officer pulls a term sheet out of a folder and passes it to you across the table. You look it over. You ask some questions of your own, even some "stupid" ones. Their answers reveal how they're thinking.

Entr'acte

Movement 3 has an entr'acte. The curtain comes down on the first part of the movement, and while the music continues you and the other dancers are offstage. You use the time first to discuss the term sheet with a trusted counselor. Then you go talk with the folks in the other ballet you've been dancing in when not dancing in this one. (Yes, you should make a point of dancing in multiple ballets.) You compare the term sheet you just got with the term sheet the dancers in the other ballet had given you yesterday. By the time the entr'acte is over, you've prepared your answer.

The grand finale? (and beyond)

The curtain rises, and you are again across the table from the chief innovation officer and the executive. You present your decision about their "offer." A discussion ensues. The curtain closes on the ballet.

What happens after the curtain closes? There are only two possible outcomes: deal or no deal. It depends on your decision. You may say "No, thank you" to the executives' offer. You may accept the term sheet as is or as it emerges from the further discussion that ensues.

Let's assume that you decline the offer. What went wrong? Maybe you're not interested in the idea the CIO and the executive proposed you build together, or you don't value the next steps in the same way. Perhaps the terms of the offer were bad for you. Perhaps the term sheet is a reality check; you realize through dancing that you do not like your own idea as much as you thought you did. (I have seen that happen.) Perhaps it's in the other ballet that you dance a *pas de deux*.

Whatever your decision, you have learned.

Charting a space of opportunity

It may not seem immediately apparent, but what I have asked you to do is piece together a vision of what you will accomplish at that larger scale in much the same way I asked you to make your problem tangible. This time, though, you are making the problem tangible for others who are not yet committing fully to your innovation prototype. That is why you need to build a *story*—and that story starts from the end. In other words, it starts from the imaginary organization you envision will solve the problem you now have reproduced at a small scale, and proceeds backward from there to find in the innovation prototype a demonstration of what's possible: a proof of concept.

This is not unlike what Pólya talks about for problem solving: Take the time to explain to yourself the roadmap to a solution at scale—make a plan. Your plan to implement the solution to the problem at the right scale and your plan to conceive the solution ought to be one and the same.

Throughout this chapter, I have illustrated advocacy using the example of seeking investment because it is reasonably straightforward compared to all the other contexts. It is the advocacy equivalent of teaching elementary mechanics with the inclined plane. As you progress, you'll have to advocate to buyers, partners, strategic partners, suppliers, and so on. Advocating to investors is the rarest of the circumstances in which you'll have to advocate—that is, it will happen only a few times, at least by comparison with how often you'll advocate to customers. Still, the logic is about the same in every advocacy circumstance.

Your plan to implement the solution to the problem at the right scale and your plan to conceive the solution ought to be one and the same.

No matter how pretty your innovation prototype, or even if you have a product, be sure not to fool yourself into believing that succeeding at advocacy is the end. It might be a beginning; irrespective of whether it is, your objective is to chart a space of opportunity.

Takeaways

• You can stress over raising funds or you can make advocacy come naturally through practice. When you advocate for resources for your innovating, you are fundamentally inviting others to buy into a larger reality you'll build together, and to take what you have built thus far as a demonstration that the larger reality is doable—a proof of concept.

• Many entrepreneurial recipes for advocacy present a rather static picture of a process that begins and ends with the pitch for investment. In reality, the "pitch" evolves constantly; the techniques described in this book are useful for gathering data with which to "accessorize" your pitch.

• Think of advocacy as an extension of inquiry and verification. Advocacy may first become obvious in the context of an investment deal but may continue beyond that and may also happen when you develop a partnership, as you acquire customers (whether consumers or organizations), and in general every time you find it necessary to show someone the value of joining forces.

• The reasons for advocating evolve. As you systematize your innovating, what you have learned through advocacy may inform how you build a business development function, a marketing department, or something else. The general progression of advocacy, though, remains the same: Get someone's attention, get to know one another and build trust as you unfold a proposal, and discuss specifics so the collaboration can begin.

• Advocacy helps you connect your innovating to concepts such as business model, value proposition, comparables, and competitors, and others.

• There is a way for you to turn advocacy into a habit: You just need to practice. Make every occasion an opportunity to practice.

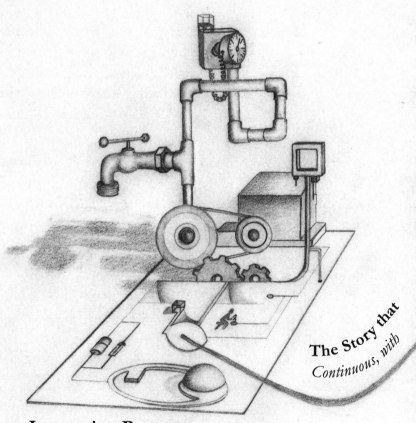

The Story that
Continuous, with

Your Innovation Prototype
*A working demonstration of the organization
and the impact of your innovating at scale*

A Destination

The larger reality you envision

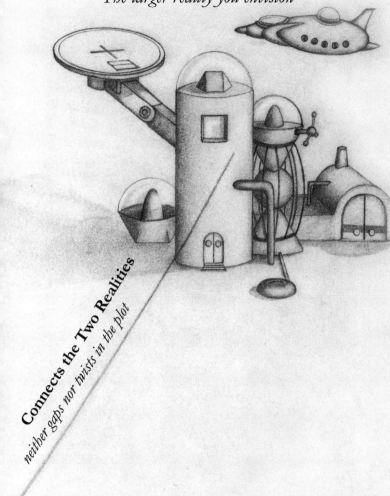

Connects the Two Realities

neither gaps nor twists in the plot

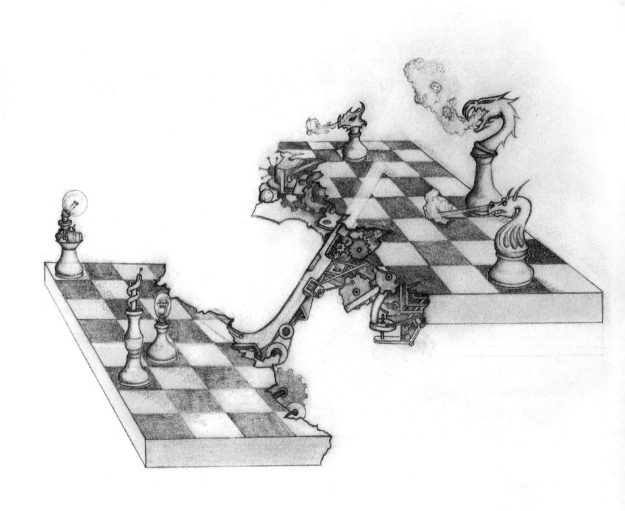

RISK, DOING, LEARNING,
AND UNCERTAINTY

If there are any risks at all as you are innovating, they're surely not there at the outset.

Just assembling some parts and talking to some people carries with it, at most, an opportunity cost of not having talked to some other people or assembled some other parts. It follows that if risks arise along the way, it's because of something else you do or are about to do. Should you do things differently, risks—if they do arise—may be different.

Even things that may seem out of your control are risks only because of choices you made that exposed you to those externalities.

You can do something different or make other choices and shape those risks.

Because your actions and choices are the reasons risks arise, you have far more control than you probably think you have. You chose those risks; you are actually the *master* of risks.

This is not the way the concept of risk is typically introduced.

Risk is usually defined as the potential for exposure to danger, harm, or loss. But that lacks the resolution you need to distinguish between the actions of your innovating and their outcomes.

Let's resolve the resolution problem.

As I write in chapter 1, at the beginning, all you have is uncertainty—and uncertainty is what fuels your innovating. Risks emerge from your failure to reduce uncertainty, which is failure to learn, thus creating the potential for danger, harm, or loss.

You are the
master
of risks.

The concept of risk appears often in the context of innovation and entrepreneurship: "entrepreneurs are risk takers"; "entrepreneurship is all about reducing risk"; "tolerance for failure"; "successful entrepreneurs thrive because they have an unusually low tolerance for failure." I take this diversity of opinions as suggesting that risk may mean different things depending on whether you intend to use the notion to guide your innovating—the doer's perspective—or whether you are assessing something someone else will do.

Risk—the doer's perspective

From a doer's perspective, there is something paradoxical about the idea of risk: Had you chosen to do nothing, there would be no risk at all. Does this mean that risks appear precisely because you set out to increase certainty about a problem systematically?

It would seem that the very notion of risk emerges only from the need to systematize your learning so your innovating can scale up. If that's the case, there has to be a way to think about risk that is more utilitarian.

I have no hard evidence to support the existence of this paradox other than a few observations from my experience teaching, doing, and advising. When I ask students and aspiring innovators to itemize risks, they don't produce a list of things that may go wrong; instead they produce a list of things they don't yet know. When asked to produce a list of uncertainties, or a list of the certainties they require to progress, they do not produce a list of actions to address what they don't know, but a list of the categories of information they are missing. Yet when the same people are presented with a technology and asked what they might use it for, their first instinct is to generate a prolific list of all the reasons it will never work for anything they can think of—and the list tends to be extensive in reasons as well as in scope of applications. Each of their answers is wrong—they answer a question other than the one I posed. But when their answers are examined together it turns out that they've produced all the correct answers they need to begin to plan their learning—as innovating—and make decisions about risks and uncertainties. All it takes is to show them how to sort the content of their answers so the initial questions are actually answered.

When you are innovating, drawing a distinction between certainty and risk that helps guide your doing can be tricky. Mathematicians, scientists, and engineers might appreciate the subtleties by analogy with "transforms." Risk and certainty may be related to one another by some kind of transformation not unlike how we use the Fourier transform to move back and forth between time and frequency domains. You operate in whichever domain makes the calculations easiest, and you communicate in whichever domain makes connection with the real-world phenomenon you are discussing easiest. Time and frequency, like risk and certainty, are just the domains in which you work.

Uncertainty and Learning

We can define uncertainty by building on the analogy with a Fourier transform, in which uncertainty is just a by-product of the fact that frequency and time (or, if you are a physicist, position and momentum) are not independent of one another. There is a limit to the precision with which you can know both. Put another way, you cannot know both at the same time with infinite precision. That defines uncertainty.

Risk and certainty are distinct variables, yet one can only hope that they will somehow be related. For instance, you may be able to mitigate risks by acquiring a set of certainties. If the analogy with the Fourier transform holds, you might expect there to be some uncertainty principle at play: the more you seek precision on the risks to which you are exposed, the wider a net you must cast to acquire the certainties you need, and thus the broader the knowledge you need and so the more you need to learn.

I have no proof for this, but the analogy seems intuitively correct. The right kind of learning contributes to reduce the uncertainty you must tolerate because it increases the precision with which you get to define risks and certainties.

For readers not familiar with the Fourier transform, let me introduce the distinction with a thought experiment. Assume that you have built *one* of something—your "innovation"—and that you decide to put it out there to see what happens.

If you have not made any promises or taken any money from investors or customers, the risk is low: at worst, embarrassment. The "dreadfulness" of the consequences grows with the scale at which you have made promises and taken resources. But risk in and of itself tells you only what will happen when you fail to do whatever is needed. Presaging that dreadfulness can supply your inquiry with the kind of existential anguish I write about in chapter 7—so you make it your objective to learn ahead of time. Then you can choose to focus on certainties or risks.

> Risk in and of itself tells you only what will happen when you fail to do whatever is needed.

You could easily post your "*one* of something" to a crowdfunding platform and see whether there is any interest. Sooner or later, someone may want to use it or buy it; worse yet, there may be multiple people who want it, but you only have one. Your biggest risk is failing to deliver, which is actually the biggest risk every crowdfunded project faces (and which has been the source of a lot of distress).[1]

One path forward is to mitigate the risk of failing to deliver by looking at what you don't yet know. You have produced only one; you now need to learn how to produce many. You can look back on the one you *did* produce and use it as a template to figure out how to produce more like it with less effort—as the founders of Greenpeace seem to have done. You can specify a set of tasks that need doing. For each task, you can assess the range of acceptable outcomes (tolerance) and the range of outcomes you know you can deliver on (precision), and close the gap by acquiring new skills and capabilities. Together, all tasks inform the range of variation your innovating may need to tolerate. On this path, you chose to learn about a risk, and in so doing you determined the certainties you needed to acquire.

Alternatively, you could decide that you really want to learn something else and make the risk of not delivering disappear altogether. Instead of learning how to produce many, you could decide to use the one you've produced to do something else. You may not even care that much about that *one* beyond how it helps you do something else. You might decide to exhibit it, auction it off, or let someone else produce and market it, but building more of it is not your focus or interest. That's what Charles and Ray Eames seem to have done with their Kazam! machine, which they created to help them implement a design in wood they couldn't otherwise make happen. On this path, you choose what you want to learn and, in so doing, you determine the nature of the risks to which you expose yourself.

The Kazam! Machine*

The Kazam! machine emerged out of Charles and Ray Eames's intuition that it ought to be possible to mold plywood into any shape easily. They built the machine out of wood and a bicycle pump.

The Kazam! machine could have become the basis for a business that mass produced bent plywood. It could have been intellectual property related to manufacturing. The Eameses could easily have opted to work on perfecting the manufacturing technique and improving their machine. Instead, the machine enabled them to win and fulfill a number of mass manufacturing and design contracts, including from the US Navy. Later, it became the central instrument they used to experiment with shape, and it enabled them to develop their iconic furniture.

All in all, for the Eameses the Kazam! machine's impact was more that it enabled them to experiment and create than anything they got out of perfecting and commercializing the machine itself. They preferred to learn with it and ended up using it to produce new designs for splints and chairs and elements for exhibits, and indirectly to develop long-lasting relationships with manufacturers such as the noted furniture designer Herman Miller.

*For more on the Kazam! machine, see http://design.designmuseum.org/design/charles-ray -eames and http://www.loc.gov/loc/lcib/9905/eames.html.

You could apply the same logic to technologies, music, industrial innovations, and social causes, to name a few domains, and arrive at the same conclusion: From a doer's perspective, risk is not an absolute concept but a relative one. There is a trade-off between what you are interested in learning and the risks you are interested in exposing yourself to. It is your job to figure out what that trade-off implies for your innovating.

This illustrates my point that you, as a doer, are better served by a definition of risk that helps you see risk and certainty relative to your innovating as ancillary variables you use to guide the choices you make—all the while learning as you scale up your innovating.

Risk can be relative. As with uncertainty, it's just a variable doers use. Both change as you learn how to scale up your innovating and make choices.

Utilitarian definitions of risk, certainty, uncertainty, and learning

As with risk, there also tends to be confusion about the terms "certainty" and "uncertainty." Most people badly want to equate certainty with predictability, but doing so makes things harder for them. These terms don't actually work the way most people wish they did—something that becomes clear when we consider their antonyms. The opposite of "certainty" is not "uncertainty" but "doubt," and the opposite of "uncertainty' is "predictability." That makes "certainty" and "predictability" not synonymous. The fact that the two terms are at most partially related implies that you can achieve certainty without having to predict everything. This should actually make things easier.

Risk, certainty, uncertainty, and learning serve very different purposes in your innovating. At a high level, risk creates the existential anguish you need to interrogate your innovation prototype meaningfully. Striving for certainties is what helps you determine what to do, what to acquire, and what you need to learn. Learning reduces your tolerance for uncertainty; it is what you do to modulate risk and certainty. In practice, this means that you'll incur some costs to guarantee survival as you prepare for some eventualities, and that you'll incur other costs to make your endeavor more robust through learning. In a sense, it's what the astronaut analogy in chapter 7 aims to help you do with your innovation prototype. However, in the astronaut analogy, the predictably unforgiving nature of space makes it possible to use the laws of physics to work backward from the one risk—dying—that trumps all others.[2]

Absent natural laws that prescribe human behavior in the face of innovations-in-the-making with the same certainty with which we can use physical laws to predict cause and effect, it is up to you to make decisions and trade-offs between certainty and risk. In so doing, you shape the boundaries of the problem you solve.

When you put all this together, it becomes easier to come up with definitions of risk, certainty, and uncertainty that are more specific than the general ones at the beginning of this chapter—in other words, definitions that are more utilitarian. We can define them in relation to your innovating and to what you need to do to persevere in your inquiries about a problem.

You can choose to define risk as a probability measure associated with the outcomes of an event. It is the least you can hope for to understand how a "system" will react. That is the best you can do if you are external to that system or if you have only a superficial understanding of how it might work—for instance, if you have no information about a system other than its inputs and outputs, or if you have nothing but analogies, comparables, or some prior beliefs with which to explain the system. (When it comes to painting a picture of risk, the comparables you use work much as parts do.)

You can choose to define certainty as a range of possibility. When you are embedded in a system and have the ability to make decisions that affect the system, you can generally outline ranges for parameters and the associated cause-and-effect relationships, and you can work with those parameters and ranges to learn to make your system more robust.

Because your job is to get a hunch out of your head and into the world, you cannot think of yourself as fully internal or fully external to a system. You need to learn to operate with both.

This is how you use these definitions. You use risk to:

• Identify dreadful outcomes that motivate your interrogation into the problem, as in chapter 7. These outcomes may come from observations, your imagination, or insights you derive from people.

• Summarize to others—investors, for example—what you'll need to learn as you scale up. This is a communication strategy to help others view your innovating in terms of your knowledge of the outcomes that would have you fail, and how prepared you are.

• Inventory and revisit things you have decided not to control for directly through your innovating. Such an inventory includes everything external to you—elements in a value chain, for instance—and must include all the assumptions you have made. (An assumption, is, after all, something you chose to take for granted and thus place outside your control.)

You use a desire for certainty to:

• Specify the experiments and actions you need to do—that is, specify the new "truths" you need to acquire to counter the dreadful consequences of risk you imagine. These specifications emerge from the process of questions in chapter 7, and specifically from your responses to the several variants of the "what might go wrong?" question motivated by your risks (as above).

• Summarize for yourself the inventory of truths and ranges of possibility that are available to you in the space of opportunity you are building. These constitute the sequence of partial proofs of concepts you've built through your innovating.

• Determine how you can make parts work together and what additional knowledge you require, and ultimately decide how to recombine parts and insights to gain control over the space of opportunity you have created.

Put simply, from a doer's perspective, risk and certainty are just variables with which you work. They help you define the boundaries of what you do. What is or is not under your control is something you are solving for as you innovate. You account for as risk all things external to your innovating, and you strive for certainty about all things internal to your innovating. (This implies that everything you decide to quantify with a probability, you have implicitly made external.) You can use the imagery supplied to you by the landscape of risks you create through your decisions to introduce that existential anguish you need to interrogate your problem and arrive at new certainties, as discussed in chapter 7. (Have I made the point about your questions needing some existential anguish?)

You learn as you progress in your endeavor. At each new scale, your tolerance for uncertainty changes, which forces you to make decisions that trade off risks to which you want to be exposed and certainties you require to move forward.

Risk—the traditional perspective

You can reconcile the doer's perspective of risk with the more traditional perspective if you think of risk as a tool that helps you communicate the robustness with which you've endowed your project.

When innovating, we tend to characterize as risks everything external to us. We also typically assess risks using our backgrounds, comparables, and how likely it is that those risks will affect our purpose. When presented with the innovating of others, we tend to characterize their endeavors by the risks they pose to us. So, whenever you find yourself on the other side of advocacy—that is, when others are advocating to you—you may equate risk with your own assessment of how likely you believe it is that the advocates will be able to use the resources they are asking for to accomplish what they set out to do. In making that assessment, you will use your background, those comparables, and ultimately your experience to judge others' ability to deliver. A significant part of that assessment will be based on whether you believe in the logic of what they propose, and on how that logic will help you believe risk is being reduced.

If you are advocating, it follows that, in addition to speaking about how inspiring your idea is, you will have to structure your story around what has been demonstrated so far and the actions you'll do next to demonstrate even more:

- The more tangible the demonstration of what you do have and know, the higher the perceived value of what you have achieved thus far and are bound to contribute going forward.

- The more specific and believable your next steps, the lower the perceived risk of investing in you.

- The clearer the distinction between your current knowns and unknowns, and how you will use resources to address the latter, the lower the perceived risk of investing in you.

- The closer the analogies—the comparables you use—to what you envision accomplishing, the lower the perception of risk.

How to determine what to account for as risk and which certainties to pursue

What you prepare to react to as a risk and what certainties you need to bring into the fold of your innovating depend largely on your tolerance for uncertainty and how quickly you think you can react to what you see.

In a web startup, for instance, reaction time is as a rule of thumb nearly immediate. At the earliest scales in your innovation prototyping, reaction can also be quite immediate. For nearly every other type of endeavor, reaction time is far from immediate. So, reaction time is connected with how quickly you can rearrange the parts of your innovating to learn something new at your current scale.

Similarly, your tolerance for uncertainty may be high, but the rest of your team, your investors, and/or your customers may think differently. So, tolerance for uncertainty is largely a function of the people you've accrued through your innovating. That means there is no formula for arriving at an optimum way to allocate your efforts between pursuing certainties and preparing for risk.

There is, however, a shortcut to minimizing risk. It requires your idea be about a product—even if you have no real certainty about it. Here's the shortcut: Constrain your endeavor to the Web, an app, or any other domain in which you need only well-understood technologies that someone else has already commoditized for use at scale (e.g., databases, web retail, 3D printable designs, Web 3.0); focus your endeavor in an archetype of a person; produce something quickly—say, a minimum viable product; don't think too much; set yourself up to monitor every activity around your product and survey everyone; and entrust your innovating to your ability to react quickly to whatever you observe.

The shortcut works mostly when scale can be trivially associated with the number of people who buy something and when everything else you might need can be specified for contract manufacturing, mass distribution through someone else's channels, or, in the case of data, can be deployed in a standard data center. In other words, it works when scaling up mostly means creating a specification and passing it on to a third party that already has a platform at the right scale. Then your risk strategy is to hedge the size of each batch of orders and manage your brand.

However, you are not limited to the shortcut. You can innovate on challenging problems, new technologies, and persistent social challenges with the same ease. All it takes is to realize that scaling up simply means figuring out how to systematize what you have learned through your innovating. For about the same effort it would take to monitor everyone who would fall prey to the allure of your minimum viable product, you can monitor the certainties you need and the risks you have created continually and learn to scale up your innovating.

Whether you take the shortcut or not, the equilibrium between what you ought to acquire as a certainty and what you ought to account for as risk to progress to the next scale gets defined dynamically by what you do as much as by the people with whom you interface.

How you explore that trade-off may be most easily understood as a game in the context of advocacy. I mean "game" here as in "game theory." For our purposes, all you need understand about game theory is that it is a way to think about how the sequence of actions of people in different roles leads to some final state—an equilibrium—in which everyone wins the most he or she can win. In the imaginary game below, the roles are defined by having either the utilitarian perspective of risk or the traditional perspective.

As you engage in advocacy, for instance, you are enacting a game that ends at an equilibrium defined by some trade-off between certainty and risk.

A game of certainty and risk

There are two players in this game: you, with the utilitarian perspective of risk, and the person to whom you are advocating, who has a traditional perspective. Your role is to decrease the perception that there is risk by clearly articulating what you know and what you don't know and asserting explicitly the things you'll learn as you scale up. You do that by sorting through your uncertainties in the form of next steps and by supplying the other player with clear, tangible milestones that allow for easy comparison.

Regardless of what you assert, other player's role is to identify and construe as risks events that will result in losses. That is, the other player draws on his or her experience and background with parts, people, and scale similar to what you presented to inform his or her perception about two things: how likely you are to learn all you need to learn given what you propose doing—that's a probability measure—and whether he or she believes your logic effectively translates into the problem being solved in an interesting way. Here interest is measured by return on investment, social impact, or whatever.

Your objective is to help the other player understand the reality you are proposing to build together, aided by the tangibility of your prototype and by comparables used as parts. The other player's objective is to assess whether the problem you can solve together is worth the risk he would have to assume. He does this by identifying and pointing out key risks that challenge your best guesses on learning, logic, and how interesting the problem might be.

As you reach equilibrium, a trade-off between certainty and risk emerges. That equilibrium defines the new problem you and the other player would solve together, what can possibly be learned, the value of that learning, and how that learning might increase the robustness of your venture. (Of course, your new problem might imply changes to costs, resources, learning, tolerance for uncertainty, and what needs to be controlled by your innovating.)

If the price you pay for the other player's perception of risk is acceptable to you and the cost to reduce uncertainty is acceptable to him or her, you've made a deal.

Practicing

As simple as the game may be to play, how to elucidate risks and certainties through conversation in a way that may be described as a game in the game-theory sense is not obvious. I am asking you to walk into a process with an initial proposal and with an understanding that the actions that will follow striking a deal do not emerge from that proposal but from the sequence of conversations. Your initial proposal only helped initiate the process.

Whether the final outcome differs substantially from what you initially proposed doesn't matter. The game is not about your initial proposal; it is about what you might build together with the other player—and that's what both parties learn through the process. The outcome may involve changes in the proposal or in how certain aspects of the proposal are perceived. Making a deal involves more than just trading chips; there is also learning and relationship building.

What's important about the outcome is that if the game is played *fairly*, the strategy that emerges at equilibrium is the one in which everybody *wins* the most given the players involved and their objectives. That, incidentally, is the only way to bring a forward-looking perspective on the meaning of "fair deal" and "win–win" to bear. And it is how the value of the deal gets defined.

To be sure, depending on the players, "winning" may mean walking away. Even then, you walk away having learned a lot. The time you spent trying to elucidate a fairer trade-off of certainties and risks with the other players is time you effectively spent changing how you view what you proposed. There are lessons in there—new insights. Just as you do after interfacing with people (see chapter 4), you may incorporate some of those insights into a new initial proposal when you decide to start the game again with a new set of players.

Making the most of the game requires two things from you: that you walk into the game prepared and that you understand how to define that set of trade-offs through conversation.

Learning to elucidate value through that kind of a conversation—that is, as the outcome of a game—is so important a skill that it deserves a way to be practiced. In my classes, I give assignments that call for my students to practice just that. I call them Karate Kid Tasks, in the spirit of the first *Karate Kid* movie. You may recall that Mr. Miyagi puts Daniel-*san* to work doing several odious tasks. Daniel has come to Mr. Miyagi to learn karate, and cannot fathom how "wax on, wax off" or "sand the floor" or "paint the fence" or "paint the house" has anything to do with that.

Karate Kid Task #1

Karate Kid Task #1 is called "That's Not What That's Worth."

The task applies to many non-obvious contexts. Its objective isn't just to strike a deal, but rather to determine the meaning of value jointly with someone else. Sometimes value is determined by the attributes of what is being proposed; at other times, it is determined by the effort it took to get a deal, measured in steps taken or in time expended. Often it is a combination of the two ways of determining value.

First I ask students to go out and try to acquire something for a price other than what's marked. They may do this at a store, or broaden the meaning of price and negotiate a school assignment, or something else. The key is to choose to do it in a setting where they perceive it will be impossible. Afterward, they share a one-paragraph description of their experience with their peers.

Their final deliverable is a one-page description of what they attempted to do, the outcome, and what caught their attention about what other students did. I use that as the basis to compose a lecture from their own experiences that ends with a discussion of the many contexts in which negotiation applies, negotiation strategies, how it all links to existing references on negotiation, attitude, and how practice is useful even when the initial objective is not accomplished.

At the conclusion of the lecture, I ask the students to consider this an assignment for life: to practice what they just learned—that is, continue to practice arriving at better and different deals continually, using their everyday lives as the training ground. When the students, later in life, find themselves in the middle of a negotiation, they will know intuitively how to define value through the process.

Translating your innovation prototype into risks and certainties

The questions in chapter 7 prepare you to use your innovation prototype to explore the relationships between certainties and risk as a function of scale. The logic set forth in chapter 10 prepares you to explore the same relationships working backward from the large-scale problem as you plan your next steps and build milestones for growth. This chapter prepares you translate back and forth between chapters 7 and 10. Learning is related to your tolerance for uncertainty: The more you learn, the lower your tolerance. You may consider your evolving understanding of risks and the certainties you need to pursue as an indication of progress.

The somewhat unorthodox mind-set I introduce using the astronaut analogy in chapter 7 prepares you for what is typically referred to as de-risking. With that mind-set, you de-risk by assuming that:

- whatever you are told is impossible can be made to work. In fact, it might actually provide a competitive advantage, but only if you figure out how to make the "impossible" actually work.

- whatever seems reasonable or valid to you now cannot possibly work. Something is wrong about it, and it is up to you to find it and fix it.

By focusing on how you are wrong and trying to correct for it, you are already as risk-conservative as it gets, and so is your innovation prototype, no matter how seemingly preposterous the underlying idea. That is, risk has nothing to do with how preposterous a problem you focused on, but with how much you've let your idea charm you into not learning anything.

Risk has nothing to do with how preposterous
a problem you focused, but with how much you've
let your idea charm you into not learning anything.

After you've worked through questions like those in chapter 7, your innovation prototype contains a set of certainties, a logical chain of possibility, a set of problems that span a space of opportunity, and a set of proposed actions you can prioritize according to how they demonstrate what's interesting about the space of opportunity that emerged from your hunch. You've done the legwork to span the space of opportunity with several distinct problems that can be solved and that you know how to vary; moreover, every action you propose to take next helps you advance in several of the problems that span the space of opportunity. Not only did you follow a risk-conservative approach to arrive here, but you've also kept open a diverse portfolio of problems that share similar proofs of concept. As far as traditional strategies to mitigate risk go, you are covered: you are risk-aware and diversified. You achieved this without sacrificing ambition. You need only translate your prototype into a story of risks and certainties. Here's how.

First, your objective is for your innovation prototype to aid your presentations of risks in two ways. One is to demonstrate how certain things related to your problem are possible. That's the value you contribute going into the conversation. The other is that your innovation prototype ought to make it easy to build a story that illustrates that what you propose doing next is beyond *possible* and is actually *probable*, meaning the odds ought to be at least somewhat better than 50-50. That's why the experimental logic of validation would have led you astray.

Second, you use your landscape of opportunity. The problems that span the space of opportunity you've identified are actually a map of several ways in which one might achieve impact. The problems convey a diversified opportunity. The logical chain of possibility is the compass for navigating the space you have mapped with actions. It can help you articulate to others the dangers and risks that lurk in the space, as well as the certainties you need to overcome those risks.

Together, the actions you propose taking next, the logical chain of possibility, and the unknowns that remain suggest a learning progression. You could probably draw that progression: one node per action, each node with multiple possible outcomes, and the logical chain of possibilities explaining how to connect the dots. Each of the problems that span the space of opportunity maps into a region of your drawing; one or more of those problems shares each node. The degree to which the problems overlap indicates how the degree to which the opportunity is diversified.

Your job isn't necessarily to share one such diagram with your conversational counterparts; it is to help others who have less information than you understand the risks and contribute their knowledge about that space. You do that by helping them see that there is a path to acquire certainties that will either ensure survival and create value or quickly show that the whole idea is not worth pursuing.

The outcome of that conversation is clarity about what needs to be learned so the project can scale. Your job is to persuade yourself there is at least one, probably many, paths to scale, and to help others see that you walk those paths building on the same principle that has gotten you here and led you to demonstrate what is possible—that you will learn a disproportionate amount from the resources you have, which at this point means that you'll learn to systematize your innovating so it scales.

Persuade yourself there is at least one, probably many, paths to scale. You do that by assuming that whatever you're told is impossible can be made to work, and whatever seems reasonable to you now cannot possibly work.

Takeaways

• The traditional notion of risk is a probability measure. Risk is hit or miss. You'll need to be conversant in that notion of risk when sharing your idea outside your circle; preparing for those conversations will be extremely helpful in your efforts to systematize your uncertainties.

• However, unless you think of your endeavor as gambling, that hit-or-miss notion of risk is limiting. Your innovating requires you to be more alert regarding what you can do about what you think you know and don't know. Simply listing risks is not an option; your options are doing or not doing.

• What matters now is how well you're prepared to handle uncertainty. That depends largely on whether you have banked your future on a single thing being true or whether your idea is robust and ready to survive changes.

• When you think about risks, you think about what happens when something that is external to you goes wrong and how to brace for it. When you think about certainty, you think about the specific truths you need to progress and the actions you can take to acquire those truths. You need to account for all those things. Your counterparts in an advocacy setting, for instance, will see them all as risks in the traditional way.

• Show a learning path through a space of opportunity that reduces everyone's tolerance for uncertainty, and that explains how you intend to trade off certainty and risk as you scale up so that you do not fail because of something you could have predicted and/or been prepared to remedy.

• Remember, there is no risk at the outset. When risk enters the picture is largely a matter of scale.

• Because you created that risk in the first place, you can make it a habit to ask: "What risk did I create today?" If you do not like it as a risk, either change what you did or figure out what new thing needs to be true so you can like what remains of that risk after you learn something new.

- If you absolutely must fail, make it come as a surprise to you and everyone, so everyone cherishes what he or she learns.

Risk, Doing, Learning, and Uncertainty

Risk does not need to be hit or miss.

Risk emerges from scale.

Risk and uncertainty depend on whether you have banked your future on a single thing being true or whether your idea is robust and ready to survive changes.

Prepare so that if your endeavor fails at a larger scale it comes as a surprise to all.

The uncertainty gap

Your certaintie

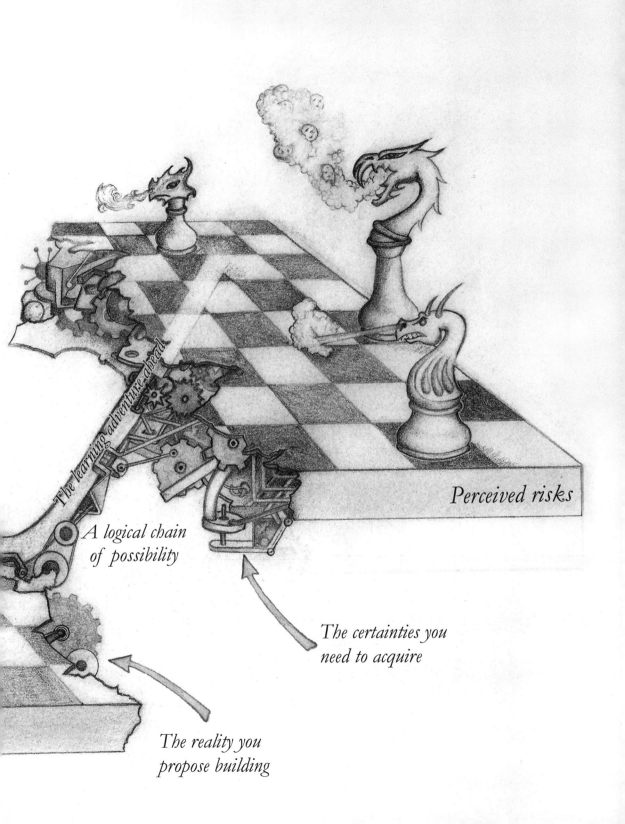

The learning adventure ahead

A logical chain
of possibility

Perceived risks

The certainties you
need to acquire

The reality you
propose building

10

SCALING UP AN ORGANIZATION

Organizations don't just grow on their own. You build them. Watering isn't enough.

After advocating for the larger reality you imagine, and getting the go-ahead, it would be only natural to think that all you have left to do is build the scaffold that supports the story arc you advocated—that is, as I illustrate in the figure that follows, place a scaffold under the curve in figure 8.1 (see chapter 8).

As it turns out, that mind-set is rather limiting and risky. It is limiting because it assumes that you have to know accurately today what will be needed when your organization is far bigger. You are leaving little room for being wrong, and by now you may have grown fond of progressing by being productively wrong. It is also limiting because it makes things such as milestones and series A, B, C investment hard to place on a timeline. And it is risky, because scale-up would then hinge mostly on the size of the market you imagined—that is, an externality.

The story arc that served you for advocacy is not a story of growth; it is a story of an imaginary and compelling future—the larger organization—that

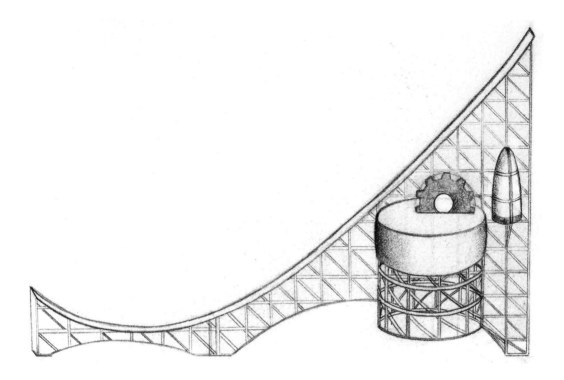

Growth as seen from a distance.

lives several scales above the innovation prototype you used to show it is possible. Growth happens in that middle you glossed over.

You build your organization by shifting your focus to that thrilling middle you had merely glossed over because it was too nuanced to make up the backbone of your advocacy story.

This chapter is about that middle. It also shows you how to use the same logic of problem solving that I have developed throughout this book to build an organization.

It follows from the examples given in the box "Growing a New Organization Atop an Old One" that thinking about building the scaffold of the larger organization you imagine might be as stressful as I note in chapter 1 with respect to finding an earth-shattering idea. You are better off thinking of your curve as supported by a scaffold that at the very least evolves over time, if not changes entirely. You are building the first scaffold.

You may take that reasoning further and adopt the position that you'll end up building multiple organizations, each one atop the previous one. That would mean that you do not even have to worry about whether a decision is optimal for the larger reality—it just needs to work today. As you build the next organization, you'll reuse parts from the old one and you'll get to implement everything you learned from the previous one. Eventually the new organization might take over what the old one did and do it more simply than you can do it now.

As you build the next organization, you'll reuse parts from the old one and you'll get to implement everything you learned from the previous one.

Incidentally, this view of growth as a sequence of increasingly larger organizations—each new, each an innovation prototype for the next—solves many of those problems that emerge as convoluted questions when you try to reason your way through innovation in hindsight. For instance, you might ask "When do I stop innovating?" The answer is *"Never."* You continue to apply the lessons of chapters 3, 4, 5, 6, and 7 *ad infinitum*, just as a physician applies what she learns throughout her career. And you continue to learn.

You might ask "Can I be a serial innovator or entrepreneur?" The answer is "Sure." Being an innovator or an entrepreneur is a profession, just as medicine is. You drop your old hunch (perhaps now an organization), much as a

Growing a New Organization Atop an Old One

It is easy to trace how Apple evolved from its Apple I into iPods and all the way to today's App Store and make sense of it—in hindsight. It is not so easy to imagine how you might have built the Apple of today from the organization it was in its earliest years. (You can read about it in a book titled *Accidental Empires*, which includes an outsider's contemporaneous account of the company's early days.*) In many ways, Apple did not just grow; rather, it transformed itself.

The same could be said for Netflix, the example with which I opened part II of the book. A 2006 *USA Today* article describes Netflix's founder, Reed Hastings, as having upended the DVD rental business and getting the company ready for an "eventual shift to downloads."** In 2008, Hastings told the *Wall Street Journal* about partnering to create a set-top box to stream movies to Internet-connected televisions.*** In 2010, *Forbes* published a piece advising investors to "short" Netflix stock, comparing the company's strategic mistakes to those of AOL. In 2011, the *New York Times* again interviewed Hastings, revisiting his mistakes, addressing the drop in share prices, and—for the first time—discussing whether Netflix competed with HBO.**** In the five intervening years from the first to last article cited here, Netflix went from competing with Blockbuster and Walmart, to being equated with the AOL failure, to competing with HBO and Amazon—admitting to having made mistakes all along. Judging by the nature of the competitors attributed to Netflix, one might argue it was in fact three different companies.

For one last example, consider Harry Schechter, CEO and founder of temperaturealert—an Internet of Things company—and a regular guest speaker in my classes, where he has been chronicling the evolution of his company. He shared with students that, after looking back on the growth spurt that had resulted in his company's growing by an order of magnitude, he felt as though he had built a new company atop the old one.

*Robert X. Cringely, Accidental Empires: How the Boys of Silicon Valley Make Their Millions, Battle Foreign Competition, and Still Can't Get a Date (Viking, 1992).

**Jim Hopkins, "'Charismatic' founder keeps Netflix adapting," *USA Today*, April 24, 2006 (http://usatoday30.usatoday.com/tech/products/services/2006-04-23-netflix-ceo_x.htm).

***Jason Riley, "Movie man," *Wall Street Journal*, February 9, 2008 (http://www.wsj.com/articles/SB120251714532955425).

****Panos Mourdoukoutas, "Can Netflix correct its strategic mistakes?" *Forbes*, October 10, 2011 (http://www.forbes.com/sites/panosmourdoukoutas/2011/10/10/can-netflix-correct-its-strategic-mistakes/#2715e4857a0b7ac0297f3bbc); Andrew Goldman, "Reed Hastings knows he messed up," *New York Times*, October 20, 2011 (http://www.nytimes.com/2011/10/23/magazine/talk-reed-hastings-knows-he-messed-up.html).

physician may move her practice elsewhere. You walk away with what you learned, and move on to the next hunch or hospital.

You might also ask "How might an incumbent innovate?" Incumbents innovate by leveraging everything that supports their current core competency—that is, what they have learned—to support their new core competency. After all, if you are that incumbent, someone was going to challenge your current core competency—sooner or later. It might as well be you.

What you need to do is not unlike what you've done to get to this point—that is, everything I've discussed is practice for what comes next. You continue to work with parts and people, you systematize, your innovating scales up, and an increasingly larger organization results. Others will see that your organization grows.

Organizations don't just grow on their own

Why is what I write in chapter 8 about the curve not enough for building an organization? It's because the curve is missing the middle. (The drawn curve does have a middle, but what it represents excludes what I will explain as the middle.)

The curve encompasses everything you've done so far and the future that you've promised—that is, the "final" destination you're aiming for—defined by a market, a self-sustaining business model, and the comparables you brought together to stitch together the semblance of an organization. All you have is an innovation prototype, a request for resources, and an enticing future that lives several scales above where you are today. That's why the shape of the curve resembles a hockey stick.

The middle is the building of your organization. It's what you're doing now. It obviously has what you've done up until today as its foundation, and it is obviously heading toward the future. But whereas in chapter 8 the curve only implies a middle, here you have to extend it.

In your advocacy, you presented an innovation prototype that illustrated at scale a vision that implied an organization with distribution channels like those of some other organization, a business model like that of Company Y, a strategy like the one employed by Company V, a battery of offerings emerging from a platform similar to the platforms those two organizations have, and customers like those of Company Z. It may also have had analytics and "Big Data" or whatever is fashionable at the time you read this—all parts. The organization your prototype implied solved a problem.

Let's say you succeeded in your advocacy and got the funding you were seeking. Then, you built an organization according to a business model different from what you had envisioned and advocated, or some variant of what you had presented. It is not as if you lied. As you advocated, you surely addressed what you would do with the money right away in the form of next steps and milestones. You sold a future, and you were given leeway to try and build it under whatever conditions exist and with what you learn along the way.

You have a problem you've solved for, and that from which you can work backward. You have an innovation prototype you've built at scale that demonstrated the large-scale problem, that is by now fairly sophisticated and ripe for systematization, and that represents an organization implicitly. It includes a problem statement redefined through advocacy and a scaffold of certainties needed and risks to anticipate. You can view your innovation prototype as a plan for how to solve the problem. In a way you have already solved the problem—except your solution is at the wrong scale, and it would be really difficult for anyone else to benefit from it now. You need to develop the solution at scale.

That is your starting point.

You scale up by discovering what about your prototype is less sensitive to scale and by finding ways to systematize what you've learned.

Growth as scaling up

In problem solving, as you prepare your solution for "prime time," you don't explain how you came up with the solution. Rather, you show others how to solve it. That requires working backward from the problem being solved, verifying your solution, and finding ways to simplify steps. You grow by scaling up in the same way.

As a first approximation, you can think of growth as a consequence of scaling up. Simply put, scale-up is about doing more of what you just did at a lower cost per unit. That's quite similar to the operating principle introduced in chapter 3: *Learn a disproportionate amount with the resources you have now.* You scale up by discovering what about your prototype is less sensitive to scale and by finding ways to systematize what you've learned. As you systematize, you are simplifying your solution. Those who will benefit from your innovating will be able to verify it as a solution to their problem only when it has become simple enough for them to use for their own purposes.

As you simplify, the prototype and everything you learn from it become new parts and insights. Systematizing your prototype allows you to transition to managing the organization that emerges from the prototype and gives you the bandwidth you need to tackle the larger unknowns you were unable to address at the previous scale.

Put another way, systematizing now mostly means trimming the "fat" from the parts and insights you've used (that is, designing parts that will do just what you need), making those parts and insights work together robustly, verifying each truth (i.e., the certainties you need to acquire), specifying everything so others can carry out your specifications, and beginning to work on identifying the unknowns you had left for now (i.e., for a later scale).

As you systematize at this scale, you'll find opportunities to start to make real some of those non-material parts described in chapter 3 that you may have only diagrammed, enacted, or emulated. You may need to build organizational functions, processes, workflows, development functions (business or otherwise), and hire people, and so on. Now is the time when some of the concepts from management about organizations, entrepreneurship, and innovation begin to acquire the increasingly specific meaning that gives them strength.

How scale-up works in manufacturing suggests a good analogy to what you need to accomplish. As with a large industrial complex, your working organization needs to ramp up to achieve the efficiency with which it will someday operate. Consider, for example, the process from discovery of a new chemical reaction to production. In the lab, a reaction occurs in several steps, executed with several different pieces of equipment, much like you work with parts in your innovation prototype. At the other end of scale-up, at production, the same reaction will occur at a radically different scale across different pieces of industrial equipment (the more standardized the better, the fewer steps the better). Even the stages of the reaction may be executed differently. There will be different environmental conditions, gauges, sensors, a control room, automation, safety measures and inspections, and regulatory compliance considerations. The organization will have a business model, do cost analysis, meet standards, attempt to make the most of each product and by-product, have environmental considerations, need transportation, and so on.

At production scale, simplifying to remove one step can make a world of difference in benefits; in the lab, it doesn't matter so much. In the lab, maximum yield can remain theoretical; in production, you produce what you produce. The difference is so stark that every single scale-up operation goes through at least one pilot phase precisely for the purpose of verifying and simplifying all the things that could only be assumed at a smaller scale.

The analogy with an industrial complex is useful for understanding what goes into conceiving a scale-up plan and the reasons why you need such a plan. Two things are particularly important. The first is the one more widely acknowledged: A progression from lab to pilot to production demonstrates the viability of the process and establishes real expectations for production. Many new challenges emerge along the way that test the initial idea and will require significant engineering, at the same time making it necessary to figure out ways to handle, recycle, and (one hopes) extract a benefit not only from the main item produced but also from all the by-products. It becomes impossible to reason about engineering and management in isolation from each other. To be sure, while a pilot plant is nowhere close in yield to a production plant, it has nothing to do with the minimum-viable-product concept. The pilot plant is fundamentally not a viable plant from an economic standpoint; it's value stems from what it demonstrates. It is a proof of concept that something is possible; it is a reproduction at scale of the larger problem you'll solve, though the scale is already larger than that of your earlier proof of concept (the lab). It can be the next innovation prototype.

Growth as a sequence of organizations that operate at different scales.

The second is generally given less consideration but is probably more important: The sequence from lab to pilot to production generally unfolds in different plants. What is carried from one plant to the next is what has been learned; the systematization of the process along this trajectory justifies the next investment in capital resources. The smaller plant may be kept for production, repurposed, or in the case of contract manufacturing left to the next client. It may be kept running to generate revenue with which to fund the development of the production plant. Most of the employees who learned with the organization move on to the next plant.

It might help you think about the organization you are building in much the same way, because it is generally easier to imagine a working organization by analogy with other organizations of similar size than to plan or even understand with the precision you need the many transitions an organization needs to undergo to achieve the "dream" scale. Still, though, you may keep on using the dream scale for inspiration. Once you learn enough with the first version of your organization, it will be easier in your next scale-up to create a second, new organization than to morph the first one—and so on as you grow. The second organization you build will "cannibalize" the first: People will carry over, and you may end up running both organizations concurrently for a while. But at some point it will probably be easier to serve the needs once addressed by the first organization with the more efficient second organization, and to dispense with the first.

You have some choices. Imagine your path to the dream scale as growing a single organization or as building successive new ones atop older ones. On the first path, you may end up stressing over how to grow the single organization to do things it was never designed to do in the first place. Considering how much learning is involved, by the time you actually know what you are supposed to do you realize how ineffective the old organization really is. You need to simplify. But because you're on the single-organization path, that means trying a reorganization—which is what large corporations opt for at this point.

On the second path, creating a new organization on top of the old one, you can salvage parts from the older organization as well as everything you've learned so the new organization is effective at that next level. The old organization partially funds the building of the next. Eventually, you figure out a way for the new organization to do what was worth keeping from what the old one did, and you can "kill off" the old one altogether.

Does the latter path—successive organizations atop older ones—work if you ever become a large corporation? If you become big enough, the new organization that gets built will cannibalize units of the larger, existing organization. IBM and some other companies seem to have been doing that for quite some time.

The history of IBM is consistent with what I explain here about innovation through the creation of entirely new companies. In *Mastering the Dynamics of Innovation*,[1] James Utterback recounts the story of IBM from typewriters to mainframes to personal computers, and points out that IBM's success seems to have hinged upon the foresight of creating a PC division with access to IBM's powerful brand and negotiating power but that was otherwise isolated from the Mainframes division it eventually cannibalized. Since 1994, when Utterback's book came out, IBM has done the same thing several more times. It sold its lucrative laptop business to Lenovo to focus on its consulting and services business, and since that happened Deep Blue evolved out of Deep Thought to become a chess master and then Watson the *Jeopardy* champion, and continues to evolve through a series of reinventions of the original idea.

The evolution of IBM is consistent with the picture of growth through creation of entirely new organizations within the old. In contrast with explicit approaches to planned obsolescence or cannibalization, the decision to create a new unit responds to a perceived opportunity—a hunch—and not to a desire to end an existing business unit. The decision to end an existing business unit seems to be made later, when the two business units can actually be compared. The new unit might also absorb what the old unit did or collaborate with it. This is what seems to be happening as of the writing of this book with the cloud services division IBM Watson, as the company accrues new "Big Data" interests such as health care.

Whatever your choice—growing through reorganization or building atop the old—what you do to scale up remains the same: identify and verify invariants (the things that remain constant at whatever scale), discover new problems, and learn what actually happens when you do it so you can further systematize. At every next scale, what you have is a new "plant."

Early on, when the scale-up step is bigger than your current organization (e.g., you jump from having twenty employees to having sixty in a very short time), describing what you are doing each time as "building a new organization" may be more accurate and a better depiction of the enormous management and organizational feat involved. You can think of growth as the organizational equivalent of a code rewrite.

Scaling up an organization as problem solving

As mentioned above, you can also think of growing an organization as problem solving. You have an innovation prototype that—like a figure in traditional problem solving—represents *all* the elements of your problem solved. Your prototype should by now include everything discussed up to chapter 7 as well as the outcome of systematizing for advocacy and risks emerging from chapters 8 and 9. In chapter 8 I discuss how to practice advocacy to reason and restate the large-scale problem with precision, and in chapter 9 I discuss how to reason about a problem from the inside out and back in to fine-tune what needs to be done in terms of certainties and risks. That raises the point that advocacy and risk assessment as discussed in chapters 8 and 9 are important components of your innovation prototype on their own, regardless of whether you seek outside investment.

You may need to work first on a related and more accessible problem, which may emerge from recombining different aspects of the logical chain of possibilities and the problems that span the space of opportunity.

A sequence of proofs of concept

The next step is to break the large problem into smaller, more accessible problems. That can be seen to mean many things. From the perspective of a solver, a more accessible problem may be any of those you produce in chapter 7 that you know are interrelated. From the perspective of an investor, a more accessible problem may mean an adjacent market. From the perspective of this book, a more accessible problem may also mean a proof of concept. All these examples really point to the same idea: Once the more accessible problem is solved, you'll be able to point to the organization that solves it as a demonstration of something; it will help you understand better the larger problem; and it will allow you to begin to systematize aspects of your future larger-scale solution. You'll also be able to use what you demonstrated as a proof of concept with future partners of what you will do next, and use aspects of what you put together as a part at the next scale.

In a sense, your smaller problem is an innovation prototype of what's to come next.

Why is this growth? The parts and people of scale

That last idea summarizes the convenience of this approach. You work only on the middle: You change what you have based on the certainties you acquire, make your organization work, and you get a new set of truths needed to progress to the next scale. If your current endeavor produces enough money, you use it to fund scale-up. If it does not, you raise more. Each time you raise more money, you're advocating. Every sale of a product comes from advocacy. Each time you advocate, you are explaining to your potential partners the big new thing you'll build together given what you have already demonstrated as *possible*, *probable*, and *doable*.

This should help you reconcile with the curse of innovation as stated in chapter 1: *It is easy to get caught up with the final result we imagine—the innovation—and believe it ought to be recognizable as such during the process.* As you begin to think about your innovation backward, from the end, you get to think of and use everything known about innovations in hindsight as parts in your scale up. In other words, you get to benefit from all that knowledge without the distraction that comes from the delusion of grandeur believed necessary to come up with an earth-shattering idea.

This may seem complicated, but it's actually rather straightforward scale-up logic: You *present* what you did (the past) to motivate where you will go (the future), but what you *work on* is that middle (the present). Most emerging organizations (commercial, social, or whatever) fail because they scale up to the future having ignored the entire present.

Working in the present: establishing next steps and milestones

The logical chain of possibility you see when you look at your innovation prototype tells you the certainties you need to acquire and their priority. Those are your paths to scale. Look at:

- the parts of your prototype. You need to make them work together in a way that requires less craft. Each requires a special design. Remove what is not needed and enhance what is needed to make it manufacturable to spec at a predictable cost.

- the unknowns you left for later. They signal ways in which you need to make parts work together so you can define the next batch of truths.

- all the things you assumed. They must be verified.

- how you acquired people. Now you need to systematize that process and help people work together.

- all the non-material parts you diagrammed or enacted. Now they must be made tangible—that is, be done either by new (material) parts or by new people.

In essence, you must walk backward from every single aspect of your prototype and simplify and systematize it so you can delegate to someone else. Your job is to manage all that while focusing on continuing to innovate so you can conquer the next scale, the next market, or the next proof of concept or demonstration—whatever name you give it.

The logical chain of possibility you see when you look at your innovation prototype tells you the certainties you need to acquire and their priority. Those are your paths to scale.

Recipe to Outline Next Steps

At this point, your innovation prototype likely supports several competing variants of the problem that gives you purpose. You may call each of these a hypothesis, an experiment, a problem, or an opportunity. They are all fairly tangible, but perhaps you wonder which one to focus on. Wondering that means you do not have enough information. You need to get that information before you can address that question.

Note that nonlinearity virtually guarantees that, no matter where you begin, you are likely to land somewhere other than where you now think you'll land. There will come a time when you will be able to make sense of why you landed where you landed. For now, you need a different way to come up with next steps other than just focusing on one of those variants.

The best way to think about next steps is not to follow any specific milestone roadmap for a single problem. Rather, gather all of the items you've identified as critical assumptions, critical certainties, and risks with which you've built the logical chain of possibility, and do the following:

1. Compare them and address the most pressing ones. Your purpose remains to seize the entire space of opportunity, not to be right about any one problem in particular.

2. Turn each into an action, which you can do by asking yourself how you can prove or demonstrate this or that with less money. If you haven't already done so, you may apply the same logic of questioning I discuss in chapter 7.

3. Rank your "needs for evidence" by how much that evidence may contribute to clarifying or dismissing any single variant. These are your "truths."

4. Several of those truths will be organizational; some will be technical. To assess them, you will have to assess the effort implied and the organizational elements you need to carry those truths forward, some of which you need to build.

Those (step 4) are your next steps. Your first milestone is the summary of what you'll demonstrate upon completing them (from step 2) and the degree of systematization you'll have achieved (from step 4).

The organization that systematizes your current innovation prototype is your first big milestone. The big experiments emerging from your logical chain of possibilities are your immediate next steps.

To be sure, you are working backward from the problem—the very one you worked hard to illustrate by way of your innovation prototype.

Doing more with less; leveraging all you practiced

So many things have to change from beginning to end, and it's harmful to assume you are already operating with the super-effective means of the future organization. That assumption makes it so much more difficult to do what you actually need to do. That is why I have presented "growth" in a way that is not typical.

This way of viewing growth also helps you see the connections across scales. If you bring the problem down to table scale, that first "downwards" section of the curve might be affordable to you. Once you have an innovation prototype, you'll have to use it to help others see the larger picture you'll be able to build together as suggested by the prototype—in other words, you'll have to engage in advocacy.

Making Organizations Work at Scale

The management literature abounds with work on how to structure for growth and on how to make organizations work. I point to further readings in the epilogue to this book. This chapter does not aim to replace any of that literature; rather, my objective is to help you see how growth relates to scale and to learning so you may use many of the skills you've sharpened through your innovating to specify the growth objectives you will then need to manage.

Your innovating prepares you to grow an organization and take advantage of the management literature on the topic of how to make organizations work and grow—although it may not seem immediately apparent.

In reality, the sophistication of parts and the roles people play change with scale, just as the problem itself changes. You can no longer gloss over certain aspects. Parts that you may have been able to get away with emulating at smaller scales (such as distribution channels or manufacturing challenges) gain tangibility. The need for specialized management knowledge increases. However, everything I've discussed so far about how to make your problem increasingly tangible with what you have remains valid at the present scale.

What changes is that with scale comes an opportunity to apply everything we know about how make an organization work effectively. What you have learned about innovating helps you stay one step ahead. It helps you make the objectives for that growth tangible, and it equips you with a means to assess whether your organization is indeed working toward the goal you imagine. It also helps you separate what you've built so far (an innovation prototype) from what you aspire to build going forward (the next problem solved). Economies of scale do not emerge from reusing or fine-tuning the product you built, but from building the next product anew—incorporating into it everything you now know and leveraging, if not repurposing, some elements of the infrastructure you built as parts.

Together, these examples illustrate the progression of growth through "rebuilding" and learning to scale up. They also suggest, contrary to commonly held beliefs, that there may not be a big difference between software companies and high-capital-intensive companies with respect to the time required to get from a first hunch to evidence. That is, restricting attention to a single product, building on largely commoditized platforms, or rushing to market may not necessarily get you there any sooner, but scaling up will. It still took about four years for both Google and Genentech to get to evidence.

If you are "bought" at some later stage, you'll also advocate for what the acquirers will be able to build next with what you have demonstrated. Even then you are advocating for a future. Whenever you are selling a product or a service, you advocate to the prospective buyers that they will be able to accomplish more with than without it; it should be the same if you are selling the entire organization.

It's the same when you discuss with potential strategic partners. You are advocating the future you'll build together on the basis of what you've demonstrated. It is *always* the same: You build a working proof of concept, and you advocate for a viable, believable, and better future.

There is a corollary to all this. By the time you've "grown" your organization, it may turn out to solve an altogether different problem than the one you spoke about in your "beauty pageant" days. The people who may have cared back then walked the path with you and influenced the evolution of the problem. They won't care whether the future you envisioned back in the day corresponds exactly to what the present has turned out to be; they are happy if it has grown. Everyone for whom you actually solved a problem will be delighted. No one else really matters.

It follows that the super-duper beautiful problem you used to build your advocacy story—the beautiful idea that got you started—no longer matters. It mattered then because it helped people believe in you and because you believed in it. Whether it ever mattered after that is hard to say.

At the time of advocacy, it is often difficult to gauge the difference [be]tween data and evidence. It takes time to arrive at the evidence needed [to] support the future vision. Typically, evidence that an opportunity is [real] emerges along the way, and it comes with learning that guides scale-up st[eps]. That is often when it becomes apparent that a "rewrite" or rebuildin[g is] needed.

For example, Sergey Brin and Larry Page reportedly developed and "t[ook] live" a search engine called Backrub at Stanford University in 1996. It [was] shut down shortly thereafter because it reportedly used too much netw[ork] capacity—that is, it had scale problems. The new search engine they cre[ated] led to Google, founded in 1998.

In the language of this book, Backrub was a first innovation prototyp[e], one that did not scale up well but was working. It helped Brin and Page l[earn] enough to build a prototype that actually scaled. It wasn't until four y[ears] later, when Google AdWords was launched with 350 customers, that a[ctual] evidence of the search engine's value to online advertising became availa[ble].

The first social network Mark Zuckerberg created, Facemash, wen[t] on Harvard University's network in 2003. Numerous difficulties invo[lving] network and organizational issues emerged, suggesting the need for a [full] rewrite. Facemash, like Backrub, was an innovation prototype. Face[book] emerged about two years later, and implementation of an ad strateg[y fol]lowed in 2008. The advertising strategy would finally produce genuin[e evi]dence of the platform's marketing potential.

Genentech, one of the world's first biological pharmaceutical comp[anies], was founded in 1976, building on academic research into recombinant [DNA]. A year later, the company produced a human protein (somatostatin) [in E.] *coli* for the first time. In subsequent years, Genentech scientists succes[sfully] cloned insulin and growth hormone. Genentech's first clinical trials [of hu]man insulin began in 1980. Both the academic research and the early G[enen]tech research that produced the first somatostatin were innovation p[roto]types. It wasn't until clinical trials commenced, four years after its foun[ding], that Genentech began to accrue real evidence of safety and efficacy [of a] biological drug.

Takeaways

• Organizations don't just grow on their own. You build them. You may end up building multiple organizations, each one atop the previous one.

• The story arc that served you for advocacy is not a story of growth; it is a story of an imaginary and compelling future—the larger organization—that lives several scales above the innovation prototype you used to show it is possible.

• It's straightforward scale-up logic: You *present* what you did (the past) to motivate where you will go (the future), but what you *work on* is the middle (the present). Most emerging organizations (commercial, social, or whatever) fail because they build for the future having ignored the entire present. But you do not have to worry about whether a decision is optimal for that rosy future—it just needs to work today.

• As you build the next organization, you'll reuse parts from the old one. You'll get to implement everything you've learned. You systematize, your innovating scales up, and an increasingly larger organization results. Eventually the new organization takes over what the old one did and does it more simply.

• Growth and scale-up work like problem solving; no one cares how you first came up with the solution. You work backward from the problem being solved, you verify your solution, and you find ways to simplify steps. You grow by scaling up in the same way: discovering what about your prototype is less sensitive to scale; finding ways to systematize what you've learned so you simplify your solution. The prototype and everything you learn from it become new parts and insights. Systematizing your prototype allows you to transition to managing the organization that emerges.

• The organization that systematizes your current innovation prototype is your first big milestone.

• Once you learn enough with the first version of your organization, it will be easier in your next scale-up to create a second, new organization than to morph the first one—and so on as you grow.

• There is a corollary to all this. By the time you've "grown" your organization, it may turn out to solve an altogether different problem than the one you spoke about in your "beauty pageant" days. No one cares.

SCALING UP AN ORGANIZATION

Organizations don't just grow, you build them. There is more to scale up than aiming at a bigger earth-shattering market.

Your growth story will be enacted by an evolving sequence of organizations

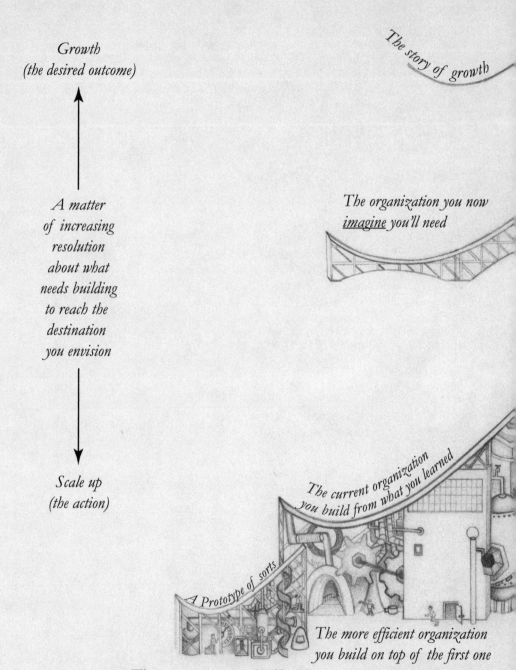

Growth
(the desired outcome)

A matter
of increasing
resolution
about what
needs building
to reach the
destination
you envision

Scale up
(the action)

The story of growth

The organization you now <u>*imagine*</u> *you'll need*

The current organization
you build from what you learned

A Prototype of sorts

The more efficient organization
you build on top of the first one

The organization you build at first

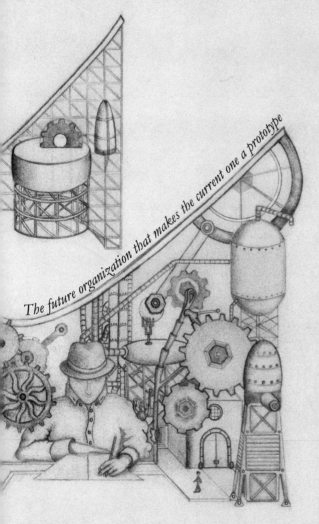

you advocated for

The future organization that makes the current one a prototype

The larger and more efficient organization you, an acquirer, a customer, or you and your partners will build on top of the previous one to accomplish even more

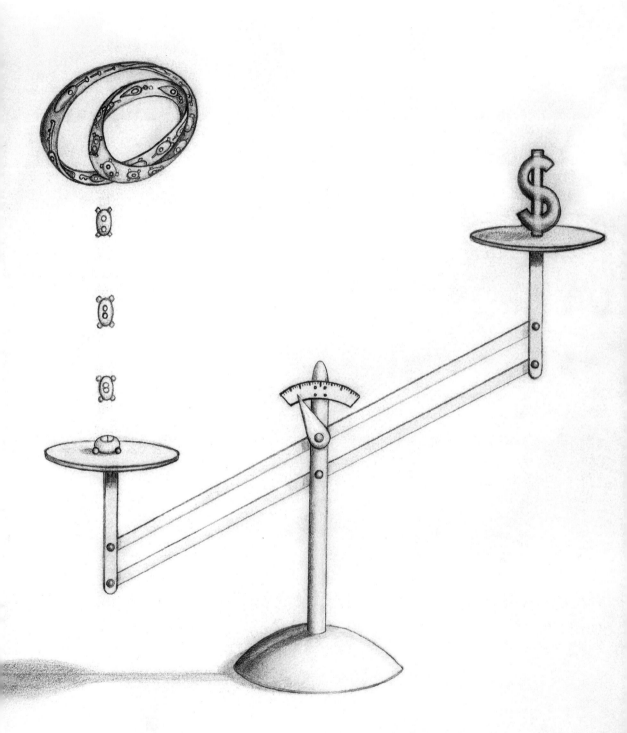

11

MANAGING INNOVATING
CONTINUOUSLY

Innovating—bringing together parts, people, and scale in a way not done before—is something you do continually; innovation is the afterthought.

You can create the conditions for innovating anywhere. In fact, you can turn innovating into a *continuous* process.

Chapter 8 introduces the growth curve as a way to connect your imagined larger reality back to your innovation prototype. The curve is an admission that serious scale up is needed, no matter how probable and doable the endpoint. Chapter 9 supplies you with strategies to ascertain certainties and risks with which to increase the resolution at which you look at the "middle" of the curve. Chapter 10 further increases the resolution of the curve and suggests that growth—particularly high growth—may not be distinguishable from layering a new organization atop the old one, in much the same way you layered proof of concept upon proof of concept in earlier chapters: by repurposing parts and insights and learning to scale. The "old" organization—no matter how big—thus becomes an innovation prototype or a set of parts you already have.

Put this way, growth can be a consequence of innovating continually, not of innovation. The question that remains, then, is how to know when to start innovating. A common answer—a stress-inducing one—is that you want to do it before a challenger makes you obsolete. If you are a startup, the common answer is "before you run out of money." If you are a research lab, the common answer is "when the technology is ready to get out of the lab." These are all truisms.

In this chapter, I offer an alternate answer: Be the challenger by having a process for innovating continually. The process helps you use everything you have now as an auxiliary part. Its purpose isn't just to create a new product, although that could be an outcome, but to help you explore continually for updates to the problem your organization solves—without concern for your current core competency. With the approach to documentation I discuss in chapter 12, the process can help you explore all other uses for everything you've already built, or currently have or do.

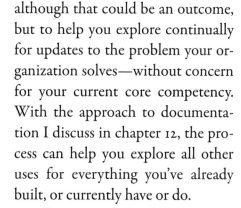

Innovating is something you do continually; innovation is the afterthought.

This is unusual in that innovating may be less about building from what you now consider to be your core competency—whether it be knowledge or product—than about accepting that you may venture out of what you now think of as your core competency. Everything you once had built to support that core competency—what you already know how to do and what you already have— can be your innovating advantage. This follows directly from my discussion in parts I and II of this book. When an innovation finally emerges, you'll be able to explain in hindsight how it evolved from your core competency; before then, your core competency—what you already know how to do, and everything that supports you in doing it—is an asset. You cannot let it become a constraint.

This chapter describes principles to implement innovating as a process as they might apply in a corporate setting. The same principles could easily be translated to a research or educational environment, or to any other environment in which a group of people decide to engage in innovating together.

You should think about innovating outside your core competency. Everything that supports that core competency today is a part with which to innovate.

At the beginning of the twenty-first century, some of the most successful venture funds began to move away from a sole concern with selecting investment opportunities and incubating ideas and toward developing internal processes to conceive and evolve ideas to create those opportunities themselves. This represents a significant shift in mind-set analogous to the one I propose in this book. It amounts to abandoning the notion that ideas are good and need only be "incubated" and instead presuming that initial ideas are just beginnings—hunches—and will have to evolve significantly before an opportunity can be conceived.

The principles of innovating as a process

Innovating is about finding a way to bring resources together to demonstrate a path to solving a real-world problem worth solving. The outcome of the process ought to be a road map to use resources for that purpose. If any new ideas emerge, it will happen along the way; at the outset, nothing needs to be new.

The objective of the process isn't to accelerate, incubate, or dress up ideas. It is to evolve whatever initial hunch you have—probably beyond recognition—until a decision about resources can be made. This is the exact opposite of how many so-called innovation labs are organized. Their process tends to involve thinking up a new idea, pulling together a slide deck and a business case, and getting on the schedule of the committee that doles out funding. The committee compares and selects ideas, and someone begins to work on them. However, as noted in chapter 1, most of the ideas that *eventually* led to lauded innovations were, at the outset, neither new nor distinguishable from failed ideas. Rather, they evolved, which seems to imply that one should avoid selecting altogether and instead emphasize the need for the initial idea to evolve.

Innovating is a different process altogether. It is premised on the notion that, whatever the original idea, the sole certainty is that it will need to evolve. It will be easier to evolve it if the focus is placed on a problem. In chapter 6, I describe some of the principles that differentiate innovating from merely culling ideas. They are: Innovating thrives when the exploration is open-ended rather than constrained in any way other than by defining an initial problem; the objective is to evolve and refine problems through the combination of accessible parts and people; ideas are neither good nor bad, and are not comparable; ideas are inconsequential as an end-product for a process; and

> Whatever your original idea, one thing is certain: It
> needs to evolve.

consequences emerge from translating resources into strategies to demonstrate value.

A few other principles apply to innovating as a process.

• You want ideas that help you extract the most value from every dollar spent. The precise dollar amount assigned as a budget is not informative. That is, you do not want to limit yourself to ideas that can be "shown" for, say, $1,000 or less.

• Thriftiness ought to be implemented by how the process is run, not by an arbitrary budget number. Parts and easy procurement are most important.

• The more people engaged in some way in the process to conceive the new idea, the broader the ownership of that idea. The process needs to enable many people to contribute in different roles. This means that the entire decision about an idea cannot be based solely on whether it accrued a team.

• Senior leadership needs to own, champion, and enact the process. Value emerges from sending a clear signal that the process is an important business unit.

• The more people in an organization empowered to contribute insight, the better teams will learn how to use as a part everything the organization currently has or does.

• The outcome of innovating isn't an idea, but a series of choices to be made about a space of opportunity about which a decision can be made.

The endpoint of the innovating process is a decision (a verification recipe)

The decision to be made is about a specific space of opportunity and the resources required to demonstrate it and seize it, as I describe in chapters 7 and 9.

By the time a group is ready to present the outcomes of its innovating to advance to the next scale, it should have coalesced as a team, have a proposal

> The decision to be made is about a specific space of opportunity and the resources required to demonstrate it and seize it.

to use resources, be able to articulate clear next steps that "de-risk" the opportunity, have a significant first milestone, and be able to make a tangible demonstration that hints at its future value and shows it is not only *probable* but *doable*. Otherwise, there is little upon which to base a decision.

Conversely, teams are ready to propose advancing to the next scale when they can articulate a problem that an organization can solve (and cares to solve). Specifically, teams ought to be able to:

- outline an organization, demonstrating the salient aspects of that organization using their innovation prototype

- identify the innovation that is needed, pointing to the specific aspects of their innovation prototype that will require a new approach (be it a new kind of gizmo or business model or social movement or whatever)

- explain how an organization can be built up incrementally, through next steps, to address the problem, the resources that will be needed, the uncertainties those resources will reduce, the space of opportunity that opens up, and the impact that ought to be expected from engaging in such an endeavor

- describe risks and certainties, pointing to the innovation prototype.

The first milestone of the proposed "road map" is, in a way, the first verification recipe. Under the definition of a verification recipe given in chapter 2, that milestone ought to make it possible to recognize a solution. That is, it ought to enable decision makers to assess whether the space of opportunity is *wrong* before more money is spent on subsequent milestones. Then a straightforward decision can be based on that verification recipe: Do you care about the problem enough? Has the team given you reason to believe that the resources it requests are the best way to discover whether the idea has merit? Are those resources well employed to get that extra clarity regarding the opportunity? Should the time of the team's members be spent on this or would it better for them to focus exclusively on their regular jobs? Did the team manage to garner support? Most of these things can be ascertained through inspection of that first milestone. The dynamics of this decision process are not unlike those discussed in chapters 8 and 9.

In contrast with the idea-harvesting processes I discuss in chapter 6, here ideas self-select, and decision makers can analyze them individually on the merits of the plan of action proposed, not their relative appeal in comparison with other ideas. Instead of having thousands of ideas in need of "funneling," asking originators to evolve ideas into decisions gives them an opportunity to work on ideas until they become plans or else drop them.

Overturning the typical way "innovators" in corporations seek and get approval for next steps diminishes the adversarial relationships that are a natural by-product of a "contest" in which alternative ideas are "pitched" to decision makers who hold the purse strings.

Seeding the process

Throughout this book, I have hinted at three options for initiating innovating as a process: as an open call seeded with a series of open-ended problems, with an innovation prototyping kit, or by asking people to identify an interesting real-world problem. In addition, the process may be readily adapted to begin with new technologies, perceived market needs, or any other input.

Select ideas on their own merits, not on their relative appeal compared to other ideas.

At its conclusion, the process may lead to an innovation prototyping kit that can be seen as a form of documentation, to an innovation prototype ripe for advocacy, or to both.

A process, a workshop, or an innovating environment can begin with a space to work, parts, people, and a desire to work on something that can scale to the real world. Participants in the process, whatever their roles, ought to be interested in solving a real-world problem.

> A process, a workshop, and, more generally, an
> innovating environment can begin with a space
> to work, parts, people, and a desire to work on
> something that can scale to the real world.

The sections that follow outline ways to engage people, work with parts, and induce a scale-up logic to accomplish the objectives discussed in the preceding sections. These suggestions emerge from my experience developing innovation programs internationally in university and corporate settings. In each implementation of this process I have carried out, I've always been able to identify elements in the culture of the organization that facilitated adaptation to its specific context.

People

Innovation prototypes accrue people as they mature, and, as discussed in chapter 6, teams coalesce only when they begin to interact with parts and interface with other people. This means that from the outset your process needs to provide ways to engage people actively in multiple roles. At the very least, your process will need to reserve an active role for people who may encounter your team's innovating through advisory or decision-making relationships, and it will need to allow for those people to come and go.

You implement three kinds of *active roles*: team members, advisors, and decision-makers.

The main considerations for forming teams around an idea are, first, that presumptive team members should be genuinely interested in the problem, possibly even having chosen it. Whether they know one another already and whether their backgrounds qualify them to participate are somewhat less important, in view of the discussion in chapters 3 and 4. Often, viewing the process as a way to advance in the organization is a good motivator.

Second, creativity need not be an attribute; it can be an outcome. Whether people think of themselves as "creative" according to whatever dictionary definition you think applies doesn't matter. Their creativity will emerge as they use their backgrounds, skills, and abilities to communicate with others to make a problem tangible.

Third, having people from different disciplines involved helps reduce disciplinary biases. Of all the benefits attributed to such groups, the only one I have observed and can take as a given is that they are much less likely than others to fall into the trap of using discipline-based jargon and standard assumptions. The variation in their backgrounds is more likely to be an asset than a constraint. Members of a cross-disciplinary team working together have no choice but to learn from one another.

Coalescing People Around Projects

The project selection process is a carefully orchestrated sequence of events designed to group participants into teams that will later work on bigger problems or projects. It is important that the participants select the project they want to work on based on the problem, not on who else is in the group.

The selection process followed in the MIT innovation workshops is instructive and can be adapted to other settings. It is staged as a sequence of exchanges of information between participants about the larger projects. (In the parlance of the university, these are called Capstone Projects: multifaceted assignments meant to serve as a culminating experience in which students take what they've learned and apply it to examine a specific idea or create something specific.)

First, participants read some prepared real-world problem statements and discuss them briefly with whoever is sitting in the next chair. They address questions such as these: "Which technologies would you build?" "Who might care?" "Can you describe the problems in your own words?" Participants then identify their top three preferences among six to twelve problems presented. Next, assembled in groups, participants discuss how they would go about solving the problems. This same discussion then ensues in a second group based on second-choice preferences. Participants are then surveyed again for their top preferences.

The theory behind the process is to simulate an economic market for ideas, and the way it is staged guarantees an information exchange in which there is first a preference-diffusion step and then a "near miss" review. The full process spreads out the distribution of participants across projects. Through this process, teams are announced and given a bit of time to come together and discuss what they learned about the projects in the information-sharing rounds.

In this way, participants form an opinion about each project based on what they understand on their own *and* what they come to understand from others through conversation. When it comes time to share insights with their newly formed teams, participants think about the task at hand and what they have learned so far.

In addition to having a process to craft teams, you need processes to allow this team to evolve throughout their innovating.

You'll also want to include a role for advisors. You may think of advisors as mentors; for instance, having the leadership of different units participate in innovating as advisors is a way to set up a mechanism for early buy-in from the broader organization. But you may also think of everyone in your organization as a potential future advisor: The members of your organization have specific expertise and skills that may become valuable for the team's innovating, and they too need to know that it is acceptable for them—and potentially rewarding as well—to share that expertise.

Mentors

Mentors may come from inside or outside an organization. They may be managers, entrepreneurs, or technologists seasoned in the application of technology to industry. Wherever they come from, they must be accustomed to providing feedback relevant to business impact much as a good member of a company's advisory board would do. This extends to providing the same kind of feedback—not technical advice—even for a somewhat crude technical prototype. Mentors are going to see a lot of odd-looking contraptions that participants will present as prototypes to explain their innovating.

In addition to those criteria, mentors should never feel any obligation to become team members.

If the outcome of your innovating process is a decision, then sooner or later you'll need to engage someone who cares about seeing the end results of your process. That is the role of decision-makers. You want decision-makers to be close to the process; that offers them an opportunity to engage sooner with innovating projects as advisors, mentors, or through intermediate checkpoints, placing less emphasis on "demo day." When emphasis is placed on innovating and not on making a presentation, the conversation is about scale-up, not approval.

Your efforts will probably benefit from having someone with some power and authority feeling ownership of the entire process—someone who can let everyone know that answering questions in the course of this process is considered part of their jobs. That's how teams discover the parts in the current organization, and also how your organization can use innovating to learn to communicate across its own silos.

Parts: the doer logic of the process

Innovating is predicated on *doing* as a way to learn by prototyping innovations. As a doer, you care less about something looking pretty or about receiving approval. You care about it working. The concept of "being wrong" boils down to this: Something is wrong because it was put together wrongly or because it was put together for the wrong people. Something works when it stops being wrong. The only way to know you are wrong (or, by extension, that you are no longer wrong) is to be told by nature ("put together wrongly" so it doesn't work, as above) or by people.

Parts provide tangibility and are the means by which teams become thrifty. However, anticipating every part any team might need can be cumbersome. Indeed, one of the jobs of a team is to identify as parts things you already do, produce, or procure. But for parts you don't already own, there has to be a means of procuring them. Endowing a team with an arbitrary budget, say $1,000, emphasizes the budget, not thriftiness as a function of the process. You can as easily implement the constraint on thriftiness in the use of resources by introducing innocuous constraints: an overall time limit; a daily time limit for prototyping; and limits to what extra resources can be requested before producing a plan—not a budget, but a constraint that specifies what can be obtained with no further approval and what will require approval, which ought to be related directly to the demonstration value of the requested resource.

In my experience, this latter constraint translates into teams self-regulating and never spending more than a couple of thousand dollars. That, in turn, trains them to think in the same thrifty way when proposing next steps. In many organizations, the constraint turns out to be easier to implement if the organization already has a procurement function.

Scale

As your teams evolve the problems they work on, they will also evolve the level of sophistication with which they put together parts and people to achieve impact. This, by the way, is the very basis upon which you will be building the foundation for an innovating environment.

The doer logic continues. Participants will need new parts and new information from people to continue prototyping their problems. As they do so, and as they scale up, they will confront the need to reason about organization. The more clearly outlined the organization, the easier it is to define the problem it solves. It is a virtuous circle.

The problem the organization will solve lives at some scale. The current proof of concept is *scale 0*. As teams get closer to defining what ought to be proven at *scale 1*, their prototypes ought to inform all that could go wrong at that scale, so that *scale 1* informs *scale 2* and each successive scale informs the next scale.

Eventually, as your arbitrary deadline approaches, teams will either be ready to produce a logical chain of possibility to advocate for more resources or will be ready to wrap up their projects in an innovation prototyping kit that documents ways in which parts and people in your organization may be brought together differently to explore new spaces of opportunities. This can become the searchable basis for your innovating knowledge base, and the way in which your organization as a whole applies the guidelines for documenting I discuss for an individual in chapter 12.

The conductor: orchestrating and managing the process

Participants in the process will need lots of guidance. In some settings, there may be considerable pressure on those who've been selected to participate. They may experience some anxiety; after all, the corporation may still expect them all to continue to do their regular jobs. And they may be banking on the process turning into an opportunity for long-term personal career growth.

The purpose of guiding is to encourage participants to find their own answers (through technology or impact prototyping) and to help them make their questions and their assertions increasingly concrete.

I have already discussed how the outcomes of innovating may hint at different ways to repurpose what an organization currently does in order to seize different spaces of opportunity. Just as people learn, so too does the organization.

For the people engaged in the process, *learning* is, in fact, the overall objective. Participants should learn how to prototype and scale up a problem again and again, and in doing so learn about the organization and about where to address questions.

The learning objectives—ones that will pay dividends in the wider innovating environment being established—are reinforced by guiding participants to adopt a can-do attitude, take ownership of their learning, and embrace the opportunity to make decisions in the face of uncertainty.

At some point, participants will have to go back to their regular jobs. Before that, though, the projects they work on need either to end or be advocated to continue, at which time projects need to deliver something *tangible*, *thorough*, and *rigorously inspiring*. In my workshop in the educational setting, these manifest, respectively, as the following:

- a prototype, a physical illustration, or a mockup of the proposed technology solution or a precise demonstration of the problem (tangible),

- a brief written report (thorough), and

- a brief presentation to an appropriate audience from outside the workshop (rigorously inspiring).

The specific elements of these final deliverables lay the foundation for how innovation prototyping will take place in your innovating environment. They may constitute the basis for documentation and possibly a searchable innovating intelligence.

For each project, the teams that propose to continue would have to prepare to walk into movement 2 of the imaginary ballet in chapter 8.

I brought together for these presentations an audience of members of the MIT ecosystem who agreed to act as a "council of wisepersons." I included academics, administrators, investors, and entrepreneurs.

In an academic setting, a presentation event is a typical way to reenact movements 2 and 3 of the imaginary ballet. In a corporate setting, the correct way to reenact the same movements is to create a meeting with actual decision makers; no role-playing is needed. If they find an interesting avenue to pursue, they will become the equivalent of members of the board of a startup.

As I discuss in chapter 6, the concepts for innovating just outlined for a corporation apply as well to research organizations—and obviously the workshop concept fits other educational settings.

Takeaways

• You can create the conditions for innovating anywhere and make it a continuous process. Innovating is about finding a way to bring resources together to demonstrate a path to solving a real-world problem. Innovation is the afterthought.

• You can build an engine for innovating that yields impact, strategic insight into new problems, and increased efficiency; "configure" it to produce new products, companies, causes, policies, inventions, innovation kits, identify talent and resources, or develop professional innovators.

In a corporation, it can be a unit tasked with finding ways to reuse everything to solve a new problem before what you do becomes obsolete.

In research/academic organizations, it can make research inclusive of other disciplines and propagate it all the way to society.

Generally, it can drive the conversation towards making societal problems tangible and solvable.

• Whether refining your organization's offering or planning its obsolescence, your organization can be an innovation prototype. That is your innovating advantage.

• Your innovating advantage comes from these principles:

Innovations begin open-ended and emerge from exploring a problem.

You may venture beyond your core competency. In hindsight you'll always be able to rationalize how your innovation evolved logically from it.

At the outset, "Ideas" hardly look new or different from failed ideas. "Ideas" are inconsequential. Your objective isn't to accelerate, incubate, or dress up ideas for show. It is to evolve them beyond recognition, until a decision about resources can be made or time runs out. That's easier done if the focus is on the problem.

Your innovating process reveals how to access parts and people in an organization.

Everything you've already built, know, and have can be a part.

Everyone in an organization can contribute as innovator, mentor, or expert. The more contributors, the broader the ownership, the greater leveraging of existing capabilities.

Arbitrary budgets for ideas are confusing. Resolve to extract the most from every dollar spent instead of limiting yourself to ideas that can be "shown and told" for, say, $1,000.

The outcome is a scale-up sequence of choices to allocate resources to obtain answers about a space of opportunity.

Innovations, if any, emerge in the scale-up. At the outset, nothing needs to be new.

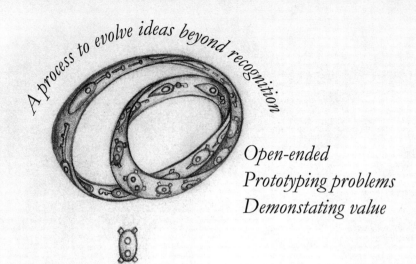

A process to evolve ideas beyond recognition

Open-ended
Prototyping problems
Demonstating value

A tangible outcome

A decision:
demonstration value resources

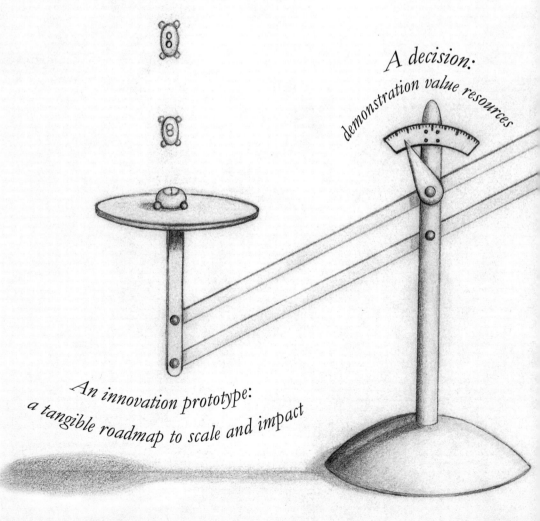

An innovation prototype:
a tangible roadmap to scale and impact

A request for resources to progress up scale

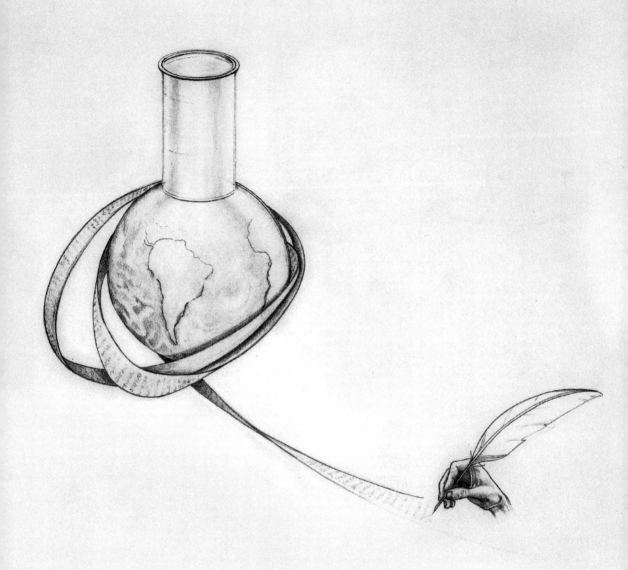

THE WORLD IS YOUR LAB, SO YOU NEED A NOTEBOOK TO CONVERSE WITH YOURSELF

Get it out of your brain before the memory of an idea you once had becomes more important than the idea ever was. Later, when your brain finds it again, it will become a near miss. That and everything it helps your brain remember can be a part.

It took only a madeleine for Marcel Proust to remember his childhood.

Might this imply that you may dispense with your note-taking software if you make a point of bringing a new muffin to every meeting? That may be stretching what Proust describes, but it helps bring up a whole bunch of important questions: What's the best way to keep a record of my innovating? What is the purpose of those records? Are they for me? Someone else? Posterity? Are they just a defensive weapon in case of litigation? Am I creating a journal? A manual? Do I need a hunch to start? Can documentation help me arrive at a hunch? *What am I even supposed to write down?!*

There are various theories about how our brains might work and what they imply about the best ways to remember and relate different pieces of information. Some say we should document everything we do, but it's unclear which approach to documentation to take. Perhaps writing things down in a notebook by hand is best. Maybe it's best to use some kind of digital device. There may be some other way that works best for you. And then there's the question of whether you should carry your notebook with you at all times, or if it's ever okay to leave your digital device behind.

Perhaps we should just follow Proust's lead and eat more muffins.

One way to answer these questions is to focus on the benefits of documentation. Some of them are easily understood in hindsight.

Consider the day your repeat innovating has led you to become insanely famous. Your sketches and prototypes may now be able to fetch unfathomable amounts of money at auction—for you or perhaps for your descendants. They may be displayed in museums, as have Picasso's sketches on napkins, John Lennon's glasses, Leonardo da Vinci's notebooks, and even the Apple I. Collectors value the ephemera generated by innovators.

You may even derive value sooner if ownership over your work is challenged in court and you have to show your records to support a patent or some other piece of intellectual property.

Perhaps documentation will provide you solace when it facilitates the reunion with an idea you thought lost. In theory, after all, the idea has been *documented*.

The benefits of documentation will be most easily understood by *future-you*—the real expert at what you are doing and fully equipped with hindsight.

This all implies noteworthy documentation that is systematic and searchable. It also implies that you ought to go about documenting diligently. It says little about what to write down or sketch, and it provides nothing in the way of how to derive value from your documentation *now*.

Personally, in the context of innovation prototyping, I am torn. It is not clear whether having notes and records about everything you do, observe, and hear is, in and of itself, help or hindrance. Notwithstanding all the benefits I've just described of keeping detailed records and proper documentation and making it all easily accessible, is it possible that documenting everything could get in the way of just thinking about things? Or is there a way for what, when, and how you record to help you take advantage of your brain's evolutionary adaptation to relate distinct pieces of information?

How do you derive benefit from documentation when you have nothing but a hunch to go on? Or even before your hunch becomes apparent? And *when*, *what*, or *how much* you should document is unclear?

The answer is to dispense with the notion that documentation is mere record keeping and instead see it as a part. Like other parts, it does little for you when stored away. Instead, approach it mechanically. It is a way to capture things you need to be alert to, and a way to help you take part in the "conversation" that is your innovating. You can derive value from documentation in much the same way Proust discovered his childhood in a madeleine.

In chapters 3, 4, 5, and 6, I discuss how to converse with parts, people, and a team. In chapter 7, I discuss how to converse with your problem through questions. In chapter 8, I discuss how to use your innovation prototype to converse with others about the problem. In chapter 9, I discuss how to trade off certainties and risk through conversation. This chapter is about how to take advantage of documentation as a means to be part of the larger conversation your innovating has become, in the way Proust's madeleine provoked involuntary memory. Beyond a record of to-do lists and results of experiments, documentation is a way to tap into that involuntary memory as well as other features of your brain.

Long-Term Benefits of Documentation

Some long-term benefits of documentation derive from the simple notion of storage and access. Keeping records is a way to capture important, specific details about things you try and more formal experiments. Those records can help with writing or defending intellectual property (IP) and, most important, reducing IP to practice. Careful records are a source of credibility in a dispute.

The world you aspire to affect surrounds you. That world is your laboratory. Documenting can be as simple as maintaining a laboratory/engineering notebook—an innovating notebook. The notebook is where an experiment first comes to live. It is the first place in which it becomes tangible. You write and sketch in it to make your reasoning and what you anticipate will happen clear to yourself—you "front load" your thinking. The notebook is also where you record what happened, what surprised you, and how it was all wrong. *Future-you*—the real expert at what you are doing, fully equipped with hindsight—will know how to make that useful. Innovation is the afterthought.

Because the world is your lab, you do not get the luxury of always being able to set up an experiment. You should still strive to make notes of what you observe as if it were an experiment you set up—if for no other reason, so you can offload the memory and focus on thinking about how to bring parts together to incorporate it into your innovation prototype at scale. Your innovation prototype is the world at scale—or at least a version of it in which the problem cannot exist.

Annotating everything about an innovation prototype has another long-term benefit: It allows you to dismantle your innovation prototype and repurpose its parts for use at your next scale. It also facilitates going back to earlier designs or rescuing earlier thoughts. The notebook in which those thoughts live is your madeleine, a time machine. It helps you converse and correct *past-you*; browsing it allows past you to surprise you with thoughts you may not remember having and from which you derive insights—just as you derive insights from interfacing with other people.

As an innovator, you are a human with a constantly evolving innovation prototype. Your job is to choreograph parts and people and scale. The notebook is the story of all of the near misses of past you. Everything in it can be a part.

To derive benefit from documentation when you have only a hunch, you must dispense with the notion that documentation is just record keeping and see it as a part.

Involuntary memory

In addition to the wealth of information we acquire through our disciplinary training and whatever we record in our notes, our brains also store memory from all our senses and mechanical movements. Sometimes our brains make connections that would seem absurd to others but make perfect sense to us. These connections emerge from a collage of experiences unencumbered by the disciplinary boundaries we may have been trained to respect.

This sort of brain activity is just like drawing the seemingly preposterous connections I ask you in other chapters to be unafraid of making. It is also the terrain upon which you can search for hunches in your experience and in the experience of others. Your brain works the same way regardless of whether you already have a hunch.

Some people use the word "intuition" to describe this phenomenon. I avoid that word for two reasons. One is that intuition is difficult, if not impossible, to practice. The other is that, like "innovation" or "creativity," the word "intuition" describes something after the fact. "Intuition" is what we call it when there seems to be no rational explanation for how we came up with a good idea. I wonder how, in that context, it means anything other than "I made it up."

Speakers of Romance languages will appreciate that there is no real difference between "making something up" and "inventing"; for instance, in Catalan, in French, in Italian, and in Spanish the same word is used for both.

The more you just let things brew in your brain, the more likely you are to have a hunch—which is, fundamentally, an intuition. If you don't yet have a hunch, that is how you may happen upon one.

If you are willing to accept that documenting is a way to become part of the conversation, and you accept that *that* may be the sole purpose for documentation today, what you need to document becomes clearer. You need to explain to yourself the *things you make up*. In keeping with one of this book's central themes, you also need to explain to yourself how you think they are wrong, or why they resist being wrong.

To ensure your documentation remains useful in the future, you just need to make sure you are diligent about keeping notes, you add dates, and you keep some sort of index.

To ensure they are useful now, whether or not you have a hunch, your notes need to help *present-you* and *future-you* continue to "make stuff up" easily. Think of them as a laboratory journal, with your laboratory defined as everything that surrounds you, and concern yourself with recording your insights, observations, and reactions. The value of what you record will be apparent only in hindsight. Unless you are already famous or you have a court case pending, the value of what you document will mostly emerge from what the notes help you think about and the kind of brain activity they trigger. That requires that you make it a point to engage actively with your notes. *The particular madeleine is as the circumstances around which you ate it.*

This brings us back to the question of how to make the best use of documentation so it allows you to make stuff up easily. In my experience, the process begins with making the most of your brain. Let me explain with an example from the field of artificial intelligence (AI).

For years, researchers in AI and neuroscience have tried to reproduce human intelligence. The same challenge emerges again and again: Our brains integrate all kinds of stimuli, memories, and language in ways that defy simple models. Our brains record information in peculiar ways that go beyond tags, files, folders, and so on. We may remember something by smell, with a motion, or through visual stimuli. Cause and effect seem to be tricky in our brains, too. When presented with the same two phenomena repeatedly, our brains will adapt to anticipate one or the other—as did Pavlov's dogs.

To operationalize these observations, there ought to be a way for us to condition ourselves to recognize patterns, unearth memories, or connect seemingly disparate concepts with ease. For instance, a memory might be triggered involuntarily by your writing something in a way similar to how you once wrote something else, or by your laying something out in a way that is similar to something you laid out previously. In both cases, of course, you have to have done the earlier "something"—which is where the practice of "documenting" might be beneficial.

It may well be that the act of reflecting on something and writing it down in a sensory-rich environment is all you need for your brain to begin developing the kinds of connections of parts and people with which you can continue innovating.

The more you make a point of recording your thoughts, even when on the go, the more you broaden the expanse of sensory information your brain might use to "annotate" your thinking. This is where the idea of *conversing with yourself* comes in: Documenting done as I've described it activates all those areas of your brain that truly come together only when you engage in a conversation with yourself.

In other words, you have to demand more from writing something down than just storing and indexing notes.

How Our Documenting Brains Work

I am asking you to take for granted that your brain is at its best when it forces itself to build a story that connects distinct and dissimilar pieces of information and memories. Whatever you store is but a means of priming that brain function when *future you* browses what you wrote and encounters an idea again.

You may need to find a compromise between the searchability of digital notes and the tangibility of "analog" notebooks. Today's digital documentation and collaboration tools and cameras are optimized for the purpose of search and storage. When innovating, your objective is to make everything as tangible as possible.

Thoughts that can be searched are not together until you put them together somewhere—wherever browsing could lead to a chance encounter with something in your documentation that spurs the conversation with yourself.

In labs, you would be asked to make extensive use of copy/paste or even printing and gluing to make sure all thoughts that belong together appear in one place.

After you start to bring things together in one place, it is easier to understand what systematizing means: it is all about bringing closer the thoughts that must come together for a certain purpose so that it takes less browsing or searching to get to all the relevant thoughts.

What you should capture in your documentation

I've offered lots of motivation and encouragement for documenting, and a rationale for doing so. But that still leaves the question of *what* to include in your documentation. In my experience, an unassuming approach works best. Some of my suggestions are general; others are more specifically linked to things you'll do along the innovation prototyping path. Some documentation is of what you learn and do; sometimes the act of documenting provokes thought.

First, you should capture whatever you are thinking, without concern for being wrong. In the future, in hindsight, you will understand what you meant and why you were wrong—if you *were*.

When you write down a fact, also capture what it would imply for that fact to go away. Write about how you would know whether something can make that fact go away. Imagine not one but as many solutions as you can that would make the fact no longer a fact. Write those down.

You are going to be making it a routine to recombine and replace parts and insights and then imagine the larger reality they now represent while trying to find out how it will not work. Those are *trial* and *error*. Describe each trial and each error unconcerned about whether you could have predicted an error before trying. Capture in your documentation the trials of others you may observe. Write down every contradiction and paradox that emerges as you are innovating.

State, restate, and restate again the problem that gives you purpose. Write it down often. My own rule for this is to do so in fewer than 300 words each time. It does not matter whether what you write down today matches the way you stated the problem yesterday.

Each time you restate the problem in your documentation, consider writing down your answers to questions such as these: Is this even a problem? Is there a related, more accessible problem? Can I draw it? Diagram it? Enact it? Can I enumerate data? Unknowns? Conditions? How can I demonstrate the problem with what I have? Can I describe it with parts and people?

Capture your thoughts without regard for being wrong. Our brains have no problem drawing the seemingly preposterous connections that fuel innovating.

> When you write about parts, remember that they
> are important only because of what they do.

When you write about parts, remember that they are important only because of what they do. So, for each aspect of the problem you want to solve, capture in your documentation an explanation of what you want each part to do. Use analogies if that helps; they'll get you to a near miss. Describe how to make the part work for your purpose. Document how much it cost. Write down what you want it to represent. And for each part, write down how you answered this question: Can I do this with some other part?

What you already know is an asset, not a constraint. Capture what you don't know; go find someone or some resource that teaches you how to do it, and write that person's name and contact information down, or spell out what the resource is and how you might get it. If you think something is impossible, it should still be in your documentation. Write it down. Who knows? Future you may find in you that elderly scientist in Arthur C. Clarke's laws of prediction.

Keep a record of every question you think of for which you'll need to find an answer, even if you don't need that answer today. Document every assumption you make; they are important elements of your evolving "questionnaire." I caution you about questionnaires in chapter 4. Here I offer them as a tool in your documentation: They summarize your questions in one place. You develop and use them in your conversations with people but you never show them; they are not survey instruments. Humans notoriously have trouble expressing themselves in questionnaires.

You're going to be meeting a lot of people with different backgrounds, information, capabilities, and skills. Write about that. Everyone is an expert at something, so find out at what and capture that in your notes. Make a note of the circumstances under which you should contact a person again.

When you write about people, be sure to capture the mundane details: who, where, when. Don't forget to write down how each person thinks you are wrong, the potential contacts they mention, and the facts they share. That is the conversation you *had* and should capture. Whatever insights you may derive after the conversation belong in a separate place in your notes. That avoids conflating your own thoughts with the conversation you actually had; why you derived this or that insight after any given conversation is a mystery and immaterial. It also keeps your own ideas separate from the people who might have sparked your thinking.

Before and after meeting people, review your "hidden" questionnaire. Did any new questions emerge? Write them down to use the next time. Were any questions solved? Write down how. Are there assumptions to revisit? Articulate how and why, and adjust the questionnaire accordingly.

As you innovate, you'll build an organization—a machine made up of people and parts. It will need management and it will need technology. Before that, though, you have to find a way reduce everything about it to everyday words. Write it down. Write down how its parts will interface with one another, and what can and cannot be systematized. Describe the basic organizing principles. I find it helps to force myself to replace every piece of jargon I might be tempted to use with something I could explain to a friend who is an expert at something else.

When you skim through your notes, think of everything you read as a near miss. Are there ways to make the near misses related? Could you somehow try several of those near misses together and find out how they are wrong?

Write about the conversations you have with people. Capture your own insights separately.

When you skim through your notes, think of everything you read as a near miss.

This is not an exhaustive list.

As you are innovating and documenting becomes a regular activity, your own list will grow. You may also want to get a second copy of this book and annotate in it why you think each and every thing I say is wrong. That way, you can grow your own list. I confess this is a shameless attempt to win a bet on how many copies this book will sell.

The "Beginning to Annotate" box has some tips for getting you started.

Beginning to Annotate

There are ways to prompt annotation. Some of these take the form of questions or the form of incomplete statements to be filled in. Here are some examples:

- How is that even a problem?
- What if ...?
- If that were true, then _____.
- Now that I know how to make this work, what happens if I change _____?
- This has got to be wrong, but why? What would make it wrong?
- What that person said cannot be true. If it were, _____.

Innovating, in hindsight

What I mean by documenting is nothing other than writing down or sketching out what results from applying the principles, asking the questions, and recording the observations I have written about in the other chapters. Your future documentation is your laboratory notebook, and your lab is the world in which you intend to solve a problem.

Throughout the book I have adopted an unassuming vantage point that may seem contrarian at times. The romantic appeal of appearing to be a contrarian notwithstanding, the fact is that the nonlinear character of innovation makes it difficult to find meaning in words that will serve to describe accurately a future larger reality when you are still at a stage at which nothing is supposed to be working.

What was the value proposition of that first laser again? The one done as a side project, with a flash lamp and a material—ruby—that countless experts had discarded as a poor fit to build lasers?

The unassuming vantage point helped you venture into the impossible. Retelling yourself your story in hindsight helps you systematize value.

But still, there is value to the story of your innovating in hindsight. It does make sense. And retelling it the way you understand it now is as important to your systematizing and to your documenting as taking the unassuming vantage point was for venturing into the impossible. That's how you eventually learn. In fact, your story in hindsight will look a lot like a recipe—one someone might follow to get to your "innovation" more efficiently, even though that recipe may have nothing to do with what you actually did when you were innovating. Let me show you, using this book as an example. The story I have told you about in this book reads like this, working backward, in hindsight:

> Get it out of your brain before the memory of an idea you once had becomes more important than the idea ever was. Innovating—bringing together parts, people, and scale in a way that no one has done before—is something you do continually. Organizations don't just grow on their own. You build them. Watering isn't enough. You may take the extreme position that you'll end up building multiple organizations each on top of the previous one. Once you build the next organization, you'll reuse parts from the old one and you'll get to implement everything you learned on the previous one to do things more simply the next time around.
>
> Together, the actions you propose taking next, the logical chain of possibility, and the unknowns that remain suggest a learning progression. It may seem that taking the scale of your innovating up a notch is the only path left on which to make progress. You don't really know whether that's true. But there's only one way to find out: Approach people who have the resources you need and propose that they join you.
>
> At some point the elderly scientist from Arthur C. Clarke's first law of prediction will concede that you've conjured your way around the "impossible."

This is all taken verbatim from the beginnings of the chapters in part III, starting with this chapter and working backward. The other chapters do not matter in hindsight because the "beginnings of an innovation are a mismatch for the stories of the organizations and innovations they ultimately empower" (from chapter 1) and because, like the stories in chapter 1, yours will "appear to be fraught with near misses and learning" that make everything look like an anecdote. What matters is that "an innovation came to [everyone's] attention because it had an impact. Specifically, [you] solved a real-world problem" (from chapter 2).

Documentation can be your first innovation prototype. Writing down a seemingly preposterous idea can get you started innovating.

———————————

Future-you will know how everything was wrong and will tell the correct story _in hindsight_

Your Lab

The world in which future-you solves a problem

The record of future-you's

Get it out of your brain
before the memory of an idea becomes
more important than the idea ever was.

near misses

YOU CAN NOW START. This is loop 2.

If your innovation prototype exists already, this is how you start again. Go to chapter 1 with a larger problem to materialize. The parts are different, the people change, and the accuracy of the conversations you have with them change at the new scale at which you address the problem. What you do stays the same.

Nothing changes except scale. Innovating stays the same. Your innovation emerges in the recurrence.

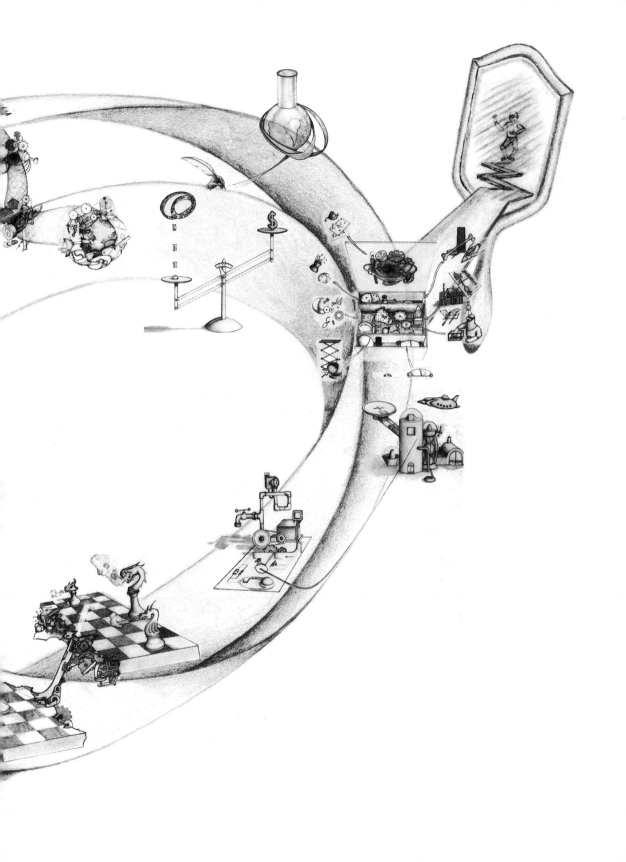

EPILOGUE: ACADEMIC COMMENTARY

Innovating is a discipline and practice in its own right that benefits from many healthy connections to several other disciplines, much as physics and engineering relate to each other and much as behavioral science relates to economics and management.

This book brings together ideas from several disciplines to explain innovating by its actions. In this commentary, you will find academic references and relevant literature across several disciplines. Some support key ideas discussed in the book or portray the opposing ideas; others help to situate the book's content in a broader discussion.

To use the language I have introduced in this book: This book can be considered a fully working prototype of a conceptual and cross-disciplinary rendering of the subject of innovating. The concepts discussed in this academic commentary are the parts with which I have built this prototype.

Bounded rationality and behavioral decision making

Throughout this book, I outline a systematic process and illustrate its roots in logic, but because the focus is on "doing" I omit references to cognitive traps and to the implications in decision making. However, the presentation of skills and capabilities is consistent with and at times inspired by Daniel

Kahneman's pioneering work on behavioral decision making, which builds on the concept of bounded rationality, and by subsequent work on cognitive traps. That is, the presentation of the methodology is logical, but I have taken care not to imply that the decision making that follows can be executed by following a similarly logical, formulaic, or *Über*-rational approach. On the contrary, I often allude to the dangers of overly prescriptive recipes and to the need to broaden the space of opportunity rather than rushing to make choices.

The book incorporates strategies for avoiding cognitive traps in a subtle way. For instance, the presentation of innovating as combining prototyping of an entire innovation with reasoning about how the problem is wrong is consistent with strategies for avoiding availability and confirmation bias presented by Kahneman and others. Because the book is intended not to show how to avoid such biases in decision making but rather to help innovators accrue skills and capabilities for innovating, I chose to introduce the concepts through an analogy with Arthur C. Clarke's laws of prediction and to not expand on the nature of those biases in the text, preferring instead to offer some suggested readings here.

Readings

Bazerman, Max H., and Don A. Moore. *Judgment in Managerial Decision Making*, seventh edition. Wiley, 2008.

Kahneman, Daniel. "Maps of bounded rationality: Psychology for behavioral economics." *American Economic Review* 93, no. 5 (2003): 1449–1475.

Kahneman, Daniel. "A perspective on judgment and choice: Mapping bounded rationality." *American Psychologist* 58, no. 9 (2003): 697–720.

Kahneman, Daniel. *Thinking, Fast and Slow*. Farrar, Straus & Giroux, 2011.

Bridging with artificial intelligence

This book's use of methods and strategies from artificial intelligence (AI) as a sort of "micro-foundation" provides some protection from the biases that emerge from the literature on bounded rationality and its effects on decision making.

Artificial intelligence has, for several years, operated at the boundary of logical rationality and empirically observable human rationality, with the objective of reproducing the latter. The Turing test is an early example of the challenges associated with operating at that boundary..

Indeed, the most mathematical branch of AI has sought to equip itself with myriad tools for fighting biases empirically. In computational learning, these biases account for what is known as overfitting and/or confusion of correlation and causation.

Artificial intelligence has also sought to create computational ways to acquire knowledge by making sense of empirical observations, with or without preexisting models. In a way, AI has created a logic that runs parallel to the empirical biases documented in the literature on bounded rationality, not because it claims to solve them but because it is concerned with the tools that help develop evidence from observables as a search problem.

The interested reader may want to draw connections between the methods outlined in this book and the logical basis outlined by Patrick Henry Winston and by Stuart Russell and Peter Norvig for knowledge representation, forward–backward logic, null space, near misses, and so on.

Using the logical tools of artificial intelligence to formulate the problem of innovation provides the micro-foundation I believe innovation needs to bridge in and out of the "black box" often alluded to by economists (see "Learning and innovating" and "No thing is new" in this epilogue).

Readings

Mitchell, Tom M. "Generalization as search." *Artificial intelligence* 18, no. 2 (1982): 203–222.

Pearl, Judea. *Causality: Models, Reasoning, and Inference*, second edition. Cambridge University Press, 2009.

Pearl, Judea. *Probabilistic Reasoning in Intelligent Systems: Networks of Plausible Inference*. Morgan Kaufmann, 1988.

Poggio, Tomaso, and Federico Girosi. "Networks for approximation and learning." *Proceedings of the IEEE* 78, no. 9 (1990): 1481–1947.

Russell, Stuart J., and Peter Norvig. *Artificial Intelligence: A Modern Approach*. Prentice-Hall, 1995.

Turing, Alan M. "Computing machinery and intelligence," *Mind* 59, no. 236 (1950): 433–460.

Winston, Patrick Henry. *Artificial Intelligence*, second edition. Addison-Wesley, 1984.

Winston, Patrick Henry. "Learning and reasoning by analogy." *Communications of the ACM* 23, no. 12 (1980): 689–703. A version with details is available as "Learning and Reasoning by Analogy: the Details" (Memo 520, MIT Artificial Intelligence Laboratory, April 1979).

Winston, Patrick Henry. "Learning new principles from precedents and exercises." *Artificial Intelligence* 19, no. 3 (1982): 321–350. For a more detailed version, see Learning New Principles from Precedent Details, MIT Artificial Intelligence Laboratory Memo No. 632, May 1981.

Winston, Patrick Henry. Learning Structural Descriptions from Examples. PhD thesis, Massachusetts Institute of Technology, 1970. A shortened version can be found in *The Psychology of Computer Vision*, ed. P. Winston (McGraw-Hill, 1975).

Iterations and the induction argument

The arguments presented in chapter 2 conceal a formal definition of iteration and an induction argument. These and the precise definition of "problem" I introduce borrow from definitions in theory of computation.

The formal induction and iteration in chapter 2 permit a definition of learning as a sum of skills that may be practiced (each of the actions in the iteration) and as an evolution toward an increasingly clear problem (the outcome of each iteration). The iteration is formally defined because it ends with the exact same structure that serves as an input to the next iteration—namely a problem statement following our three arguments. Nothing prevents the reader from arriving at multiple new problem statements, however; as a matter of fact, that is a common output from the process. These two characteristics differentiate the recurrence proposed in chapter 2 from other approaches that have been taken to associate iterating with innovation.

The induction argument concealed in chapter 2 is important to help readers understand that, although the endpoint may not be in sight, and although there are no guarantees of converging on a solution before they give up, what they are working on has the same structure as the endpoint, and the process they are following will not distract their attention—that is, if they indeed find a problem worth solving through innovation, it will become apparent through the very same steps they are using to discover it.

The complete induction argument goes as follows: You can verify by yourself that the following three conditions are indeed true at the end and persuade yourself by induction that you are best served by working on your hunch to make them apparent also at the outset:

Endpoint. The day an innovation is recognized, members of a community will be using the outcome of your innovating to address a need that was until then unmet and most likely also either unknown to them or elusive. That need, no matter how big or small, reflects something the members of a community would like to accomplish but cannot. That is the problem. Their adoption of your innovation for a given purpose implies the recognition of the existence of the problem, and their deriving a measurable benefit from it implies that your innovation does indeed solve the problem (decidable). Your delivering this innovation sustainably implies trivially the existence of a solution. This is all true the day your innovation is recognized as such: A real-world problem exists, though it might have been elusive at first; by using your solution, the members of a community verify that your innovation is indeed a solution; and your innovation renders both the problem and the solution tangible.

Step before endpoint. Look at the very last iteration of your organization or artifact before its adoption by its community. The things you can actually do are: bring parts together in a different way, develop aspects of your organization (which is another way to bring parts together in a different way), or attract new people and skills. The result of your actions may lead to a refined understanding of the problem, but the problem itself did not change—which means that at that time, you had to be able to describe the problem by means of your candidate solution, what the solution had to achieve, and your best guess as to how others would appreciate that your candidate solution solved the problem. So, in your last iteration, you must have had answers for all three conditions as well.

Formal induction. Because at each iteration what you do is to operate with parts, the organization, and interact with people, the above reasoning applies throughout all the steps of the iteration all the way to the beginning.

Some readers may find this formal induction a bit too formal; after all, the steps are not mathematical per se, and thus while the argument follows a clear mathematical logic it is really a logically structured guess.

Although the actions in the iteration I propose follow from combining George Pólya's approach with a more general and formal definition of the problem and from notions emerging from theory of computation and the definition of problem therein, the iteration that results has elements in common with other methodologies that were developed for different purposes: W. Edwards Deming's Plan-Do-Check-Act circle and all the subsequent evolution of ideas refining in TQM and 6Sigma. All methodologies share a call to action and refinement through verification of some measure of quality. The most salient difference is the object of that quality: Deming's Plan-Do-Check-Act and subsequent evolutions of the method emphasize procedural quality, whereas in this book I look at learning as the measure of quality. As an innovation prototype scales up and learning becomes more specialized, I suspect the similarities between both approaches may increase.

Readings

Deming, W. Edwards. *Out of the Crisis*. Center for Advanced Engineering Study, Massachusetts Institute of Technology, 1986.

Morrow, David R., and Anthony Weston. *A Workbook for Arguments: A Complete Course in Critical Thinking*. Hackett, 2015.

Pzydek, Thomas, and Paul Keller. *The Six Sigma Handbook*, fourth edition. McGraw-Hill, 2014.

Sipser, Michael. *Introduction to the Theory of Computation*, second edition. Course Technology, 2006.

Kits, DIY, and experimentation

The use of kits is widespread as an educational tool to help anyone understand basic scientific principles, and—in self-reliant communities such as Do-It-Yourself communities—to offer hobbyists an opportunity to self-assemble certain products. In an article for *Make*, Michael Schrage provides an overview of the long history of kits used by experimenters. Whether for hobbyists or education, the end goal is clear and well documented in the instructions that accompany such kits.

In this book, I propose a new consideration of how kits can be used: bringing the familiar aspects of kits to the domain of translational research and innovation education, where the road map to impact is open-ended. This capitalizes on the surging trend in providing resources for individuals and organizations to become technology and knowledge self-reliant: do-it-yourself and maker movements (see, for instance, make:magazine at make-zine.com) and its foreseeable relevance in academic research and education; the hackerspace movement; learning science at home with educational kits; and learning on your own through open-source platforms. These movements are often confused with open innovation or open science and similar principled approaches to aggregate the creative but "faceless" work of many, and coexist with the trend to make online higher education broadly available and discover new streams of revenue by volume through massive open online courses (MOOCs). This trend toward self-reliance may be interpreted as society's experiments with new approaches to acquire knowledge, distribute ideas, attain impact, and leverage new distribution channels to bring ideas and innovations to scale more efficiently and broadly than traditional academic channels. The method I propose in this book, specifically the processes to learn to acquire knowledge on demand, are highly synergistic with these trends.

Eric von Hippel's work on democratizing innovation and on user toolkits outlines emerging opportunities for grassroots innovation from users and innovation communities, and the implications for existing organizations. This pairs well with innovation prototyping, which offers skills for individual innovators to capitalize on the trends von Hippel has documented in his articles and books and demonstrated in his work with corporations. In essence, our books share the belief that placing kits designed for trial-and-error and experimentation in the hands of users can dramatically alter the effectiveness of the innovation process. The innovation prototyping kits I propose take this idea a step further by associating kits with open-ended problems rather than constraining them to a specific product or service type. Also, the notion of experimentation and prototyping I advocate for in this book and describe in some of my work goes beyond tinkering with technology: I use the phrase "tinker with impact" to acknowledge the need to prototype as much of the vehicle to impact as possible to iterate through different choices for users, market, mode of impact, audience.

In general, the management literature has addressed a similar kind of experimentation with a different focus on business and entrepreneurship (Schrage's *Serious Play*) and more generally as an instrument in management that organizations ought to embrace (Thomke's *Experimentation Matters*).

Readings

Biggs, John. "Raspberry Pi, a Computer Tinkerer's Dream." *New York Times*, January 30, 2013.

Bornstein, David. "Open Education for a Global Economy." *New York Times*, July 11, 2012.

Lin, Thomas. "Cracking Open the Scientific Process." *New York Times*, January 16, 2012.

Mycroft, Alan. Raspberry Pi—Why, What, How. Presentation, University of Cambridge, June 15, 2012 (http://www.cl.cam.ac.uk/~am21/slides/CAS12.pdf).

Nielsen, Michael. "Is scientific publishing about to be disrupted?" Blog post, June 29, 2009 (http://michaelnielsen.org/blog/is-scientific-publishing-about-to-be-disrupted).

Nielsen, Michael. *Reinventing Discovery: The New Era of Networked Science.* Princeton University Press, 2011.

O'Leary, Amy. "Worries over Defense Dept. money for 'hackerspaces.'" *New York Times*, October 5, 2012.

Pearce, Joshua M. "Building research equipment with free, open-source hardware." *Science* 337, no. 6100 (2012): 1303–1304.

Perez-Breva, Luis. "Commoditizing technology innovation." In Proceedings of Skoltech Innovation Symposium, fall 2014.

Perez-Breva, Luis. Prototyping Technological Innovations—Tinkering, reasoning, and experimenting: Innovation is a process. MIT Sloan Experts, October 2, 2012 (http://mitsloanexperts.mit.edu/luis-perez-breva-prototyping -technology-innovations-tinkering-reasoning-and-experimenting-innovation -is-a-process/).

Schrage, Michael. "Kits and revolutions." *Make*, special issue, January 2011.

Schrage, Michael. *Serious Play: How the World's Best Companies Simulate to Innovate*. Harvard Business School Press, 2000.

Thomke, Stefan H. *Experimentation Matters: Unlocking the Potential of New Technologies for Innovation*. Harvard Business School Press, 2003.

Self-reliance books (examples) and selected online resources

Baichtal, John. *Hack This: 24 Incredible Hackerspace Projects from the DIY Movement*. Que, 2012.

Barron, Natania. *Geek Mom: Projects, Tips, and Adventures for Moms and Their 21st-Century Families*. Potter Craft, 2012.

Buckley, Patrick, and Lily Binns. *The Hungry Scientist Handbook: Electric Birthday Cakes, Edible Origami, and Other DIY Projects for Techies, Tinkerers, and Foodies*. Collins Living, 2008.

Denmead, Ken. *Geek Dad: Awesomely Geeky Projects and Activities for Dads and Kids to Share*. Gotham Books, 2010.

Frauenfelder, Mark. *Made by Hand: My Adventures in the World of Do-It-Yourself.* Penguin, 2011.

Plant, Malcolm. *Understand Electronics.* Teach Yourself, 2010.

Thompson, Robert Bruce. *Illustrated Guide to Home Chemistry Experiments.* O'Reilly Media, 2008.

Thompson, Robert Bruce, and Barbara Fritchman Thompson. *Illustrated Guide to Home Biology Experiments: All Lab, No Lecture.* O'Reilly Media, 2012.

Thompson, Robert Bruce, and Barbara Fritchman Thompson. *Illustrated Guide to Home Forensic Science Experiments: All Lab, No Lecture.* O'Reilly Media, 2012.

Wohlsen, Marcus. *Biopunk: Solving Biotech's Biggest Problems in Kitchens and Garages.* Current, 2012.

VanderMeer, Jeff, and S. J. Chambers. *The Steampunk Bible: An Illustrated Guide to the World of Imaginary Airships, Corsets and Goggles, Mad Scientists, and Strange Literature.* Abrams, 2011.

The Home Scientist (http://www.thehomescientist.com/index.html)

Khan Academy (http://www.khanacademy.org)

Scitable: A Collaborative Learning Space for Science (http://www.nature.com/scitable)

Kuhn versus Popper

There are noticeable differences in the foundational understanding of the scientific method between my book and popular books that conflate innovation and entrepreneurship. Lean Startup, Design Thinking, Business Model

Generation, Blue Ocean Strategy, disciplined entrepreneurship, and rapid prototyping all seem to build their empirical foundation from Karl Popper's interpretation of the scientific method. For example, *The Lean Startup* and design thinking both adapt product design methodologies to the world of startups. *The Lean Startup* is rooted in SyncDev's approach to product design and management, and design thinking is rooted in IDEO's highly successful methodology born also out of product design.

I stand with Thomas Kuhn, who likened the Popperian experimental method to one from which no paradigm shift can emerge. Popper's approach significantly constrains the scope of innovation; it is useful only when the purpose is not to challenge existing theory (in this case, perhaps, a product) but to validate it and refine it.

There is some overlap in keywords between my book and the other works noted above that makes distinguishing between methods difficult at times. I have tried to address this keyword confusion by supplying precise definitions for keywords we share (e.g., problem, learning) and through this academic commentary.

Readings

Aulet, Bill. *Disciplined Entrepreneurship: 24 Steps to a Successful Startup*. Wiley, 2013.

Doorley, Scott, and Scott Witthoft. *Make Space: How to Set the Stage for Creative Collaboration*. Wiley, 2011.

Fried, Jason, and David Heinemeier Hansson. *Rework*. Crown Business, 2010.

Kelley, Tom. *The Art of Innovation: Lessons in Creativity from IDEO, America's Leading Design Firm*. Crown Business, 2001.

Kelley, Tom, and David Kelley. *Creative Confidence: Unleashing the Creative Potential Within Us All*. Crown Business, 2013.

Kim, W. Chan, and Renée Mauborgne. *Blue Ocean Strategy: How to Create Uncontested Market Space and Make the Competition Irrelevant.* Harvard Business Review Press, 2013.

Kuhn, Thomas S. *The Structure of Scientific Revolutions*, second enlarged edition. University of Chicago Press, 1970.

Lewin, Tamar. "Massive open online courses prove popular, if not lucrative yet." *New York Times*, January 7, 2013.

Liedtka, Jeanne, and Tim Ogilvie. *Designing for Growth: A Design Thinking Toolkit for Managers.* Columbia University Press, 2013.

Osterwalder, Alexander, and Yves Pigneur. *Business Model Generation: A Handbook for Visionaries, Game Changers, and Challengers.* Wiley, 2013.

Popper, Karl R. *The Logic of Scientific Discovery.* Hutchinson, 1959.

Ries, Eric. *The Lean Startup: How Today's Entrepreneurs Use Continuous Innovation to Create Radically Successful Businesses.* Crown Business, 2011.

Warfel, Todd Zaki. *Prototyping: A Practitioner's Guide.* Rosenfeld Media, 2009.

Wildman, Gill, and Nick Durrant. *The Politics of Prototyping.* Lulu.com, 2013.

Learning and innovating

Several authors have likened innovation to learning. In *Inside the Black Box*, Nathan Rosenberg outlines how innovation emerges from several different kinds of learning. He discusses several forms of learning; most notable is learning by using, which may lead to "disembodied knowledge" that in turn may lead to the modification of the original hardware. This is consistent with my presentation of parts and knowledge with which to inform and materialize a larger reality after systematizing learning. Rosenberg further relates "learning by using" to the early work of Eric von Hippel, which von Hippel himself later built upon for his book *Democratizing Innovation* and his work on product kits.

The kind of learning I discuss in this book builds on the relationship between innovation, user-driven innovation, and learning premised both by Rosenberg and von Hippel with respect to knowledge emerging from products. I take that relationship as a fundamental principle and extend it to any kind of knowledge that may be obtained about or with a part, whether a product or not. I build upon it to propose a method of exploration to guide that learning.

The relationship between learning and innovation is discussed frequently in the innovation and entrepreneurship literature. Peter Denning and Robert Dunham's book *The Innovator's Way* describes thoroughly the essential practices to achieve innovation and supports the view of innovation as a learning process. Denning and Dunham distinguish between "invention practices" (sensing, envisioning, and offering) and the "entrepreneurial practices" that follow them. Peter Drucker writes of a systematic entrepreneurship in connection with learning. Schein also relates learning with innovation as elements in organizational management.

In general, the economics and management literature does a comprehensive job of specifying the outcomes of innovation as a learning process. More recently, popular business books such as *The Lean Startup* and books on "design thinking" have invoked learning as an element in the process, although other than noting that learning happens they never define it explicitly.

The academic management and economics literature is more precise in the use of learning, and closer in intent to the use in this book. However, learning is generally considered an outcome of the process: learning happens. In this book, I explore strategies to fuel that learning. That is, I pay attention to the processes and experiences with respect to how they induce learning, which allows for beginning the conversation well before there is a plan for an organization or an idea for a product.

The notion of learning used in in this book takes that definition one step further by incorporating what we know about learning from Artificial Intelligence to build a micro-foundation for learning—such as the concept of near misses, knowledge representation, and more recent work on stories and the relationship between cognitive science and artificial intelligence. It also incorporates what is known in the education and psychological literature as *experiential learning* and the development of mastery experiences to develop an operational logic for the acquisition and practice of skills.

The behavioral and psychological literature has addressed the question of the acquisition of expert performance and the distinction between "talent" and "learning" at length with works on self-efficacy and deliberate practice. As an example, the literature on deliberate practice resolves the "born" or "made" question for a number of fields. Specifically, it discards the notion that particularly gifted individuals may need less time to acquire the specific skill. Rather, it suggests that particularly gifted individuals may find different motivations or work around resources differently so that they acquire a skill comparatively faster than people devoting less time to deliberate practice. But the overall practice time needed is the same.

The same findings have been extended to the acquisition of entrepreneurial intention and self-efficacy through education.

In their report, Edward Roberts and Charles Eesly show results that suggest that repeat entrepreneurship is accompanied with an increased success rate, which one may take for an indication that entrepreneurship can indeed be learned through practice.

This all suggests that the question of whether entrepreneurship and innovation can be learned is moot. The question whether it can be taught isn't really a question about an attribute of the field but about the skill of the teacher and the adequacy of the pedagogical approach chosen. In a recent paper, William Lucas, Ilia Dubinsky, and I presented results obtained over three editions of a workshop on innovating I developed in which we show that the self-efficacy of students, considered to be a predictor of entrepreneurial intent, can be greatly increased by developing educational experiences that build on the concepts in this book.

Readings

Ajzen, Icek. "Perceived behavioral control, self-efficacy, locus of control, and the theory of planned behavior." *Journal of Applied Social Psychology* 32, no. 4 (2002): 665–683.

Ajzen, Icek. "The theory of planned behavior." *Organizational Behavior and Human Decision Processes* 50, no. 2 (1991): 179–211.

Autio, Erkko, Robert H. Keeley, Magnus Klofsten, George G. C. Parker, and Michael Hay. "Entrepreneurial intent among students in Scandinavia and in the USA." *Enterprise and Innovation Management Studies* 2, no. 2 (2001): 145–160.

Bandura, Albert. *Social Foundations of Thought and Action: A Social Cognitive Theory*. Prentice-Hall, 1986.

Bandura, Albert. "Self-efficacy: toward a unifying theory of behavioral change." *Psychological Review* 84, no. 2 (1977): 191–215.

Baum, J. Robert, and Edwin A. Locke. "The relationship of entrepreneurial traits, skill, and motivation to subsequent venture growth." *Journal of Applied psychology* 89, no. 4 (2004): 587–598.

Berzak, Yevgeny, Andrei Barbu, Daniel Harari, Boris Katz, and Shimon Ullman. "Do you see what I mean?: Visual resolution of linguistic ambiguities." Paper presented at Conference on Empirical Methods in Natural Language Processing, Lisbon, 2015.

Bird, Barbara. "Implementing entrepreneurial ideas: The case for intention." *Academy of Management Review* 13, no. 3 (1988): 442–453.

Boyd, Nancy G., and George S. Vozikis. "The influence of self-efficacy on the development of entrepreneurial intentions and actions." *Entrepreneurship Theory and Practice* 18 (1994): 63–77.

Brooks, Rodney A. "Intelligence without representation." *Artificial intelligence* 47, no. 1 (1991): 139–159.

Denning, Peter J., and Robert Dunham. *The Innovator's Way: Essential Practices for Successful Innovation.* MIT Press, 2010.

Drucker, Peter F. *Innovation and Entrepreneurship: Practice and Principles.* HarperCollins, 1985.

Elfving, Jennie, Malin Brännback, and Alan Carsrud. "Toward a contextual model of entrepreneurial intentions." In *Understanding the Entrepreneurial Mind,* ed. A. L. Carsrud and M. Brännback. Springer, 2009.

Ericsson, K. Anders, Ralf T. Krampe, and Clemens Tesch-Römer. "The role of deliberate practice in the acquisition of expert performance." *Psychological Review* 100, no. 3 (1993): 363–406.

Fitzsimmons, Jason R., and Evan J. Douglas. "Interaction between feasibility and desirability in the formation of entrepreneurial intentions." *Journal of Business Venturing* 26, no. 4 (2011): 431–440.

Gershman, Samuel J., Eric J. Horvitz, and Joshua B. Tenenbaum. "Computational rationality: A converging paradigm for intelligence in brains, minds, and machines." *Science* 349, no. 6245 (2015): 273–278.

Krueger, Norris F., Michael D. Reilly, and Alan L. Carsrud. "Competing models of entrepreneurial intentions." *Journal of Business Venturing* 15, no. 5 (2000): 411–432.

Lucas, William, Luis Perez-Breva, and Ilia Dubinsky. "Assessing the transfer of entrepreneurial education from MIT to the Skoltech Institute of Science and Technology." Presented at Triple Helix International Conference, September 2014.

Moriano, Juan A., Marjan Gorgievski, Mariola Laguna, Ute Stephan, and Kiumars Zarafshani. "A cross-cultural approach to understanding entrepreneurial intention." *Journal of Career Development* 39, no. 2 (2011): 162–185.

Roberts, Edward B., and Charles E. Eesley. "Entrepreneurial impact: The role of MIT—An updated report." *Foundations and Trends in Entrepreneurship* 7, no. 1–2 (2011): 1–149.

Rosenberg, Nathan. *Inside the Black Box: Technology and Economics*, Cambridge University Press, 1982.

Schein, Edgar H. "Three cultures of management: The Key to Organizational Learning." *Sloan Management Review* 38, no. 1 (1996): 9–20.

Schulz, Laura. "The origins of inquiry: Inductive inference and exploration in early childhood." *Trends in Cognitive Sciences* 16, no. 7 (2012): 382–389.

Shapero, Albert, and Lisa Sokol, L. "Social dimensions of entrepreneurship." In *The Encyclopedia of Entrepreneurship*, ed. C. A. Kent, D. L. Sexton, and K. H. Vesper. Prentice-Hall, 1982.

Stajkovic, Alexander D., and Fred Luthans. "Self-efficacy and work-related performance: A meta-analysis." *Psychological Bulletin* 124, no. 2 (1998): 240–261.

Tattersall, Ian. "An evolutionary framework for the acquisition of symbolic cognition by Homo sapiens." *Comparative Cognition & Behavior Reviews* 3 (2008): 99–114.

Tierney, John. "What is nostalgia good for? Quite a bit, research shows." *New York Times*, July 8, 2013.

von Hippel, Eric. *Democratizing Innovation*. MIT Press, 2005.

von Hippel, Eric. "User toolkits for innovation." *Journal of Product Innovation Management* 18, no. 4 (2001): 247–257.

von Hippel, Eric, and Ralph Katz. "Shifting innovation to users via toolkits." *Management Science* 48, no. 7 (2002): 821–833.

Winston, Patrick Henry. The Genesis Story Understanding and Story Telling System: A 21st Century Step Toward Artificial Intelligence. Minds & Machines Memo No. 109, Center for Brains, Minds and Machines, 2014.

Winston, Patrick Henry. "The strong story hypothesis and the directed perception hypothesis." Presented at AAAI Fall Symposium, Menlo Park, California, 2011.

No thing is new

The notion that novelty, technological progress, and ultimately innovation emerge from the combination of mostly non-new components has been explored in the economics literature. Martin Weitzman, for instance, introduces recombinant growth to "open the black box" of technological progress and explain the emergence of growth as a function of the combination of earlier knowledge. Nathan Rosenberg's book *Inside the Black Box* also explores the relationship between technology and growth as well as the many indirect ways in which technology enters society and the broadened meaning of technology beyond science.

In likening the emergence of Greenpeace to the development of high-technology innovations, I am building on a broadened meaning of technology in which I associate the elements of Greenpeace with parts much as we would think of parts in a technological innovation. This broadened definition of technological innovation is consistent with how technological progress is generally presented in macroeconomics, but may be less immediately familiar and may appear to be at odds with how technology is generally defined in terms of science and engineering.

Readings

Rosenberg, Nathan. *Inside the Black Box: Technology and Economics.* Cambridge University Press, 1982.

Weitzman, Martin L. "Recombinant Growth," *Quarterly Journal of Economics* 113, no. 2 (1998): 331–360.

The nonlinear nature of innovation

James Utterback, Clayton Christensen, and several others have written (explicitly or implicitly) about the outcomes and consequences of the nonlinear nature of innovation. In general, the management literature on innovation—such as Christensen's book *The Innovator's Dilemma*, Henry Chesbrough's book *Open Innovation*, and Larry Downes and Paul Nunes's article "Big-bang disruption"—focus on the receiving end of innovation, when innovations pose threats to existing organizations and how these organizations may respond. These are all indirect observables of the nonlinear nature of innovation. Everett Rogers's adoption curve too can be interpreted as an indirect observable of these nonlinearities, as are S-curves.

Geoffrey Moore's chasm is another example. Moore's warning that the chasm may be a by-product of over-targeting of an early segment of the market and his invitation to explore alternate markets adjacencies or evolutions of the original idea are analogous to the approach to nonlinearities I offer in this book. The analogy, however, is difficult to draw in foresight, because the reference to a chasm implies a discontinuity that, in foresight, is indistinguishable from the prospect of failure. Moore's discussions of growth in numbers and of the progression to build a whole product, though different in approach, are consistent with the approach to scale through systematization in this book.

More generally, the literatures of management, business, and social science have made critical contributions by explaining the nature of the innovation process by its outcomes. Agents in the innovation process (capital, corporations, policy makers, universities, and so on) now have reason to expect nonlinearities and disruption from the innovation process (as Christensen and Utterback have showed), to welcome the appearance of a sudden change in direction and pivots as potentially good (Ries), and to view university ecosystems as important agents of economic growth (Roberts and Eesley). They now have a variety of models from which to choose their preferred mode of interaction with innovation and innovators (Chesbrough, von Hippel, Rothwell). This has created a highly sophisticated and unprecedented level of demand for innovations. In some of my earlier work, I hypothesized that this level of demand, coupled with the increased online availability of knowledge and parts, creates an opportunity to propose new strategies to empower innovators.

Whereas the literature cited above identifies behaviors, communities, movements, trends, and crowd-sourcing opportunities, I identify the capabilities and skills that individual innovators need in order to refine and diffuse innovations in a forward-looking manner.

Readings

Chesbrough, Henry W. "The era of open innovation." *MIT Sloan Management Review* 44, no. 3 (2003): 35–41.

Chesbrough, Henry W. *Open Innovation: The New Imperative for Creating and Profiting from Technology*. Harvard Business School Press, 2006.

Christensen, Clayton M. *The Innovator's Dilemma: When New Technologies Cause Great Firms to Fail*. Harvard Business School Press, 2013.

Downes, Larry, and Paul Nunes. "Big bang disruption." *Harvard Business Review*, March 2013: 44–56.

Moore, Geoffrey A. *Crossing the Chasm: Marketing and Selling Disruptive Products to Mainstream Customers*. HarperBusiness, 1991.

Perez-Breva, Luis. "Commoditizing technology innovation.." In Proceedings of Skoltech Innovation Symposium, Fall 2014.

Rogers, Everett M. *Diffusion of Innovations*. Free Press, 1995.

Rothwell, Roy. "Towards the fifth-generation innovation process." *International Marketing Review* 11, no. 1 (1994): 7–31.

Utterback. James M. *Mastering the Dynamics of Innovation: How Companies Can Seize Opportunities in the Face of Technological Change*. Harvard Business School Press, 1994.

Organizational theory and computation

In chapter 3, I outline a relationship between organization and theory of computation. In later chapters, I draw a similar connection between organization and a machine (in the physics meaning of the word). Although I arrive at that argument deductively in a non-standard way, the relationship between computational theory and organizational theory is not new; rather, it is an ongoing field of work, outlined, for instance, by Kathleen Carley in the 1995 inaugural issue of *Computational and Mathematical Organization Theory*.

Readings

Carley, Kathleen M. "Computational and mathematical organization theory: Perspective and directions." *Computational & Mathematical Organization Theory* 1, no. 1 (1995): 39–56.

Parts and modularity

The discussion of parts builds from the premise that innovation may emerge through learning by using and through a process of generating knowledge as a user of other knowledge (not just products). Parts and knowledge are interchangeable; this line of reasoning resonates with the management literature on modularity. Management has only recently begun to explore the concept of modularity as an instrument in organizational management. I believe the concept of modularity offers an interesting point of connection with the concepts I introduce in chapters 6, 11, and 12, and links with the discussion of how to make organizations work in chapters 8–10 and in this epilogue.

Readings

Baldwin, Carliss Y., and Kim B. Clark. *Design Rules: The Power of Modularity*. MIT Press, 2000.

Baldwin, Carliss Y., and Kim B. Clark. *Modularity in the Design of Complex Engineering Systems*. Springer, 2006.

Baldwin, Carliss Y., and Eric von Hippel. "Modeling a paradigm shift: From producer innovation to user and open collaborative innovation." *Organization Science* 22, no. 6 (2011): 1399–1417.

People and teams

My discussion of how innovation prototypes accrue people follows a logic that is consistent with Deborah Ancona and Henrik Bresman's discussion of the attributes of effective teams. However, Ancona and Bresman consider the team as the basic unit while acknowledging that teams are flexible in many ways, including membership. In the end, teams build innovations, but I consider the team as one of the (evolving) outcomes of the process rather than a basic unit of the process. By placing the emphasis on the problem the innovation prototype makes tangible and on how the innovation prototype accrues people, I emphasize the need for people to develop a shared language to influence and receive influence, which follows from the notion of cosmopolitan influentials from Merton.

With respect to teams, many others have analyzed quite effectively and praised the merits of cross-disciplinary work, but their starting point is that we are somehow products of disciplines. This book takes a default position that disciplinary boundaries are arbitrary—as exemplified by the Harvard University lecture series on science and cooking. I leave it up to the reader to justify why we should accept overspecialization as a default mode of operation.

Relevant to the issue of conversing with people, the approach in this book differs from traditional approaches to product design and development in which the existence of a product helps constrain the search space to a set of features (as in Karl Ulrich and Steven Eppinger's book *Product Design and Development*). *Innovating* does not try to replace any of those insights, but instead notes that in the absence of one such constraint it is first necessary to learn what questions are worth asking. That normally happens through unstructured and often unplanned conversations. Innovators need to learn to extract information from those conversations.

Those who are interested in how this approach connects with managing people may want to read Peter Drucker's book *The Effective Executive*, which provided some of the inspiration for my focus on people and their contributions in chapter 4.

As an endeavor matures and processes are required, there are excellent works in the literatures of business and management that explain how to develop more formal methods for interviewing and selecting people—a matter of management that is beyond the scope of this book.

Readings

Ancona, Deborah, and Henrik Bresman. *X-Teams: How to Build Teams That Lead, Innovate, and Succeed*. Harvard Business School Press, 2013.

Bock, Laszlo. *Work Rules!: Insights from Inside Google That Will Transform How You Live and Lead*. Hachette, 2015.

Borrego, Maura, and Lynita K. Newswander. "Characteristics of successful cross-disciplinary engineering education collaborations." *Journal of Engineering Education* 97, no. 2 (2008): 123–134.

Drucker, Peter F. *The Effective Executive*. Harper Collins, 1967.

Harvard School of Engineering and Applied Science. Science and Cooking. http://itunes.apple.com/us/itunes-u/science-and-cooking/id399227991?mt=10.

Merton, Robert K. *Social Theory and Social Structure*. Free Press, 1957.

Mohammed, Susan, and Brad C. Dumville. "Team mental models in a team knowledge framework: Expanding theory and measurement across disciplinary boundaries." *Journal of Organizational Behavior* 22, no. 2 (2001): 89–106.

Ulrich, Karl T., and Steven Eppinger. *Product Design and Development*, fifth edition. McGraw-Hill, 2011.

Problem solving

This book builds upon the theory on heuristics as outlined in George Pólya's *How to Solve It* and generalizes from that work to a more general kind of problem that lacks sufficient definition to fit into one of his categories of problems (find, prove, pragmatic).

The methods and operations I discuss in this book rely heavily on the notion that innovations solve problems. The concept "problem," however, is ambiguous. It is used differently in different disciplines, and it is rarely defined precisely. The three criteria I enumerate at the beginning of chapter 2—namely, solvable, recognizable, and verifiable, hence decidable—are enough to define a problem, and can easily be found to be true from the perspective of the endpoint of the eventual innovation. They also provide a connection with the literature on problem solving and, more specifically, on modern heuristics.

Pólya defines modern heuristics as a field concerned with "the process of solving problems, especially the mental operations typically useful in this process." The definition of problems by my three criteria draws on Pólya's powerful work without restricting our attention to mathematical problems, as he does.

Pólya enumerates three specific kinds of problems: problems to find, problems to prove, and practical problems. The first two are specific kinds of mathematical problems characterized by a clear and precise problem definition. The third refers to more complex and less sharply defined problems that require knowledge and concepts also less sharply defined. He argues that solving practical problems eventually requires solving several problems to find and prove, and that data in practical problems is inherently inexhaustible, which implies the need for judgment and approximations. Pólya argues that while more experience may be needed to solve practical problems, the fundamental motives and procedures of solution required to solve a practical problem are the same as those required to solve mathematical problems.

The examples of practical problems Pólya supplies focus mainly on large engineering challenges—for example, building a bridge—in which the endpoint is reasonably well known. The problems we are interested in for innovation, while eminently practical, are slightly more general than his examples. They satisfy all the attributes from practical problems and, in addition, the endpoint is also elusive.

Reading

Pólya, George. *How to Solve It.* Princeton University Press, 1945.

The relationship among scale, execution, and organization building

Chapters 8–10 of this book build a sequence around the elements often associated with execution in books on entrepreneurship: marketing your idea (or pitching), fundraising, and growth. I try to add a layer of resolution to these elements by associating them with scale, relating them to skills that may be refined through practice, and showing how that practice relates to the concepts earlier in the book.

My presentation is not intended to replace the existing academic literature on these topics, but to provide an anchor point for doers so they can relate to that literature. I consider this topic to be underexplored in the business entrepreneurial literature, where growth is often conflated with its observable—sales—and the discussion rarely goes deep enough to allow for an understanding of the organizational and technical implications of that growth.

Edgar Schein's work relating organization and learning points to a similar problem when noting the difficulty to relate operational, executive, and engineering cultures within an organization and the need to do so to help the organization learn and innovate. Schein discusses the creep that results from lack of alignment between these cultures in his section "organizations don't learn; innovations don't diffuse"—which is consistent with my opening to chapter 9 that "organizations don't just grow."

My presentation in chapters 8–10 does not try to address all the problems relating to organizational building Schein and others raise; that would be beyond the book's scope. But I do articulate how the combination of innovation prototyping and the drive for scale can help specify a means to align those three cultures.

In chapter 8 I outline the need for salesmanship and conviction; the interested reader may find Robert Cialdini's book *Influence* relevant for understanding the salesmanship involved in conveying an idea, and may find Anthony Weston's *A Rulebook for Arguments* useful for understanding how to support an argument convincingly. In chapter 9 I specify the outcomes to expect from a negotiation process, and how to view the negotiation as a means to strengthen and specify further the task at hand. Numerous books about negotiation, such as Roger Fisher, William Ury, and Bruce Patton's *Getting to Yes*, are relevant for understanding the scope and elements that make negotiation a method by which to specify next steps. In chapter 10 I introduce a new approach to thinking about organizational growth that brings together the precise definition and implications of scale emerging from engineering with the specifications of organizational growth emerging in the management literature. To the best of my knowledge, this conceptual approach to present growth and scale hasn't been explored elsewhere.

As a whole, the sequence of chapters 8–10 presents a recurrence rule as an aid to understanding scale and organizational growth in a way that is analogous to the presentation of learning and problem solving in chapters 7 and 2. Together, the story outlined in chapters 2 and 7–10 is a story of problem solving, learning, systematization, and optimization for scale that is unconstrained by premature choices and assumptions. I have left the elements involved in human decision making and creating organizations that work out of chapters 8–10 not because they are unimportant, but because they are either well explained elsewhere or the corresponding treatment would have implied a divergence from the theme of the book.

From an organizational standpoint, the sequence discussed in chapters 8–10 is consistent with the criteria offered in the literature for addressing the question of growth strategically through acquisitions and alliances (Roberts and Liu), outsourcing (Quinn), supply chain and organization (Fine), competitive relations (Brandenburger and Nalebuff), and opportunity identification (Shane).

Finally, my presentation in chapters 8–10 is consistent with how companies such as 3M and IBM have been able to combine internal innovation with innovation powered by acquisitions and sales—exits—from mature business units.

Readings

Brandenburger, Adam M., and Barry J. Nalebuff. *Co-opetition*. Doubleday Business, 1996.

Cialdini, Robert B. *Influence: Science and Practice*, fifth edition. Allyn and Bacon, 2008.

Fine, Charles H. "Are you modular or integral? Be sure your supply chain knows." *Strategy + Business* 39, no. 2 (2005): 1–8.

Fisher, Roger, William Ury, and Bruce Patton. *Getting to Yes: Negotiating Agreement Without Giving In*, second edition. Houghton Mifflin Harcourt, 1992.

Morrow, David R., and Anthony Weston. *A Workbook For Arguments: A Complete Course In Critical Thinking*. Hackett, 2015.

Quinn, James Brian. "Outsourcing innovation: The new engine of growth." *MIT Sloan Management Review* 41, no. 4 (2000): 13–28.

Roberts, Edward B., and Wenyun Kathy Liu. "Ally or acquire." *MIT Sloan Management Review* 43, no. 1 (2001): 26–34.

Shane, Scott A. *Finding Fertile Ground: Identifying Extraordinary Opportunities for New Ventures*. Wharton School Publishing, 2004.

Weston, Anthony. *A Rulebook for Arguments*, fourth edition. Hackett, 2009.

NOTES

Introduction

1. Rod Hilton, "The Star Wars Saga: Introducing Machete Order," Absolutely No Machete Juggling blog (www.nomachetejuggling.com/2011/11/11/the-star-wars-saga-suggested-viewing-order/).

Chapter 1

1. "Theodore Maiman Explains the First Ruby Laser," presentation at 2008 LaserFest (http://www.laserfest.org/lasers/video-maiman.cfm).

2. Jeff Hecht, *Beam: The Race to Make the Laser* (Oxford University Press, 2005).

3. The story of the laser in hindsight can be traced through several books. It appears in Wikipedia and a number of other places online. I have found the account of how the first laser was put together in only one place: a copy of the video of the interview with Dr. Maiman that was shown at a conference on optics in 2008. The video cited in note 1 above isn't even the original; it is a video capture of the video shown on a screen at the conference.

4. The statement (attributed to Ben Metcalfe, who later became Greenpeace's first chairman) is quoted in Rex Weyler's book *Greenpeace: How a Group of Ecologists, Journalists, and Visionaries Changed the World* (Rodale, 2004).

5. Greenpeace International, "A Chat with the First Rainbow Warriors," 1996 (http://www.greenpeace.org/international/en/about/history/founders/first-rainbow-warriors/).

6. "Steve Wozniak (1984 Interview)" (http://www.youtube.com/watch?feature=player_detailpage&v=7RZrv55B6Js).

Chapter 2

1. Greenpeace International, "A Chat with the First Rainbow Warriors."

2. Greenpeace, "Mission Statement," December 1996 (http://web.archive.org/web/19961228050925/http ://greenpeace.org/).

3. If your offering is a consumer product, by the time you are done you will probably no longer use the word "problem." Rather, you'll likely speak directly about a *need*. That's the usual shorthand for *the thing people are missing to eliminate a specific problem*. It is difficult to justify the use of the shorthand when all you have is an elusive problem. It just makes you sound as if you know what you are talking about.

4. United Nations, "UN Water—World Water Day 2013: Water Cooperation" (http://www.unwater.org/ water-cooperation-2013/water-cooperation/facts-and-figures/en/).

5. Somini Sengupta, "A cheap spying tool with a high creepy factor," *New York Times*, August 2, 2013 (http:// bits.blogs.nytimes.com/2013/08/02/a-cheap-spying-tool-with-a-high-creepy-factor/).

6. Kaspar Mossman, "Amyris Biotechnologies, Emeryville, Calif.," *Scientific American* 298, no. 1 (2008): 42–43.

7. George Pólya, *How to Solve It* (Princeton University Press, 1945).

8. Peter J. Denning, *The Innovator's Way: Essential Practices for Successful Innovation* (MIT Press, 2010).

Chapter 5

1. What I describe as nonlinear could, depending on the context, be further categorized as a chaotic system, as non-deterministic, or otherwise—a level of precision beyond the scope of the book. I ask mathematically inclined readers to indulge my simplification.

2. Wayne Grady, *Technology: A Groundwork Guide* (Groundwood, 2010), quoting Jacob Bielow.

3. I continue to be amazed that a definition of technology from the mid-1800s is so convenient for my purposes here—especially considering how fuzzy the standard definition is today. The Harvard professor's old definition captures at once prehistoric tools and the latest nano-gizmo or quantum-mechanical wizardry device. That is, it admits the latest near-magical "scary" stuff and the oldest well-known artifacts that ring like déjà vu.

Chapter 6

1. Were it proven that ore is homogeneously distributed across earth, acquiring more land would surely increase those odds. But then the challenge would no longer be prospecting. Rather, it would be to conceive a method or technology for extracting ore that is not concentrated in a single location. As with selecting ideas, you merely traded one problem for another problem (potentially a more difficult one).

Chapter 7

1. "Astronaut Chris Hadfield Brings Lessons From Space Down to Earth." (Interview, October 30, 2013). Accessed September 10, 2015. http://www.npr.org/2013/10/30/241830872/astronaut-chris-hadfield -brings-lessons-from-space-down-to-earth.

Part III

1. "Hazards of Prophecy: The Failure of the Imagination," in *Profiles of the Future: An Inquiry into the Limits of the Possible*, revised edition, Harper & Row, 1973.

Chapter 9

1. Jenna Wortham, "Success of crowdfunding puts pressure on entrepreneurs," *New York Times*, September 17, 2012. (www.nytimes.com/2012/09/18/technology/success-of-crowdfunding-puts-pressure-on-entrepreneurs. html).

2. Of course, there are some things for which you may not be able to prepare. For instance, if there are indeed extraterrestrials and they have unexpected intentions, chances are your protocol for that eventuality will not cover them all. And it would probably be pointless for you even to prepare for more than merely warning Houston of their existence, especially in view of how much more advanced than us they must be to have made it all the way here. In that case, you are on your own. I wonder if Chris Hadfield would mind sharing whether the astronauts' preparation considers that eventuality, and to what degree.

Chapter 10

1. James M. Utterback, *Mastering the Dynamics of Innovation: How Companies Can Seize Opportunities in the Face of Technological Change* (Harvard Business School Press, 1994).

INDEX

Disciplinary bias, 112, 151–152, 319. *See also* Cross-disciplinary perspective

"Disembodied knowledge," 365

Disruption
academic commentary on, 372–374
and incremental progress, 28–29, 48
and nonlinearity, 147–148

Distribution strategies, 77

DIY kits, 178–180, 359–362

Documentation, 328–345
approaches to, 331–332
benefits of, 332–333, 334–337
content of, 337–342
and hindsight, 332, 336, 339, 343–344
and "making stuff up," 322–336
and memory, 331–337

"Do-it-yourself " ethic, 178–180, 359–362

Doer's perspective, xxiii. *See also* Being productively wrong
and growth as building atop, 296–297
and mastery of risks, 262–267

Doubt, 267. *See also* Uncertainty

Downes, Larry, 372

Drucker, Peter, 376

Due diligence, 254

Dunham, Robert, 366

Eames, Charles and Ray, 265–266

Economies of scale, 302

Edison, Thomas, 149–150, 177

Eesley, Charles E., 373

Effective Executive, The, 376

"Elevator pitch," 135

Endpoint
choice of, 51, 228, 315–317, 377–378
and induction argument, 357–358
and iteration, 61

Entrepreneurship compared to innovation, xviii–xxi, 363–366

Entrepreneurship theory, xx–xxii

Eppinger, Steven, 376

"Eureka moment," 183

Evolving problems. *See* Innovating continuously; Prototyping problems

Execution, entrepreneurial, 379–381. *See also* Practicing; Scaling up organizations

Experimentation. *See also* Trial and error
and innovation prototyping kit, 178–179, 184, 187, 359–362
and scale, 54
and trial and error, 30, 179

Experimentation Matters, 361

Exploration, open-ended, 178, 184

Facebook, 182, 303

Facemash, 303

Failure, 12–17, 372
and giving up, 197–198
and negating your idea, 193–203
and nonlinearity, 148
and parts, 99, 113
as process of questions, 203–211
and risk, 262–265, 269
and scale, 194, 298
timing of, 30–31, 96, 193–196

Fear
and data, 133
of failure, 95
and innovation prototyping kit, 175–176
and lack of knowledge, 82–83
and nonlinearity, 148–150

Fine, Charles H., 380

Fisher, Roger, 380

Fleming, John, 149, 177

Fleming valves, 149

Ford, Henry, 20–21
innovations, 20–21, 24, 48, 66–67, 90, 105, 163
social impact, 20–21, 24

Formal induction argument, 357–359

Forward-looking perspective, 8–18, 275
and advocacy, 234–235, 240–245
Greenpeace, 11–14, 21
versus hindsight, 7, 11–17, 21–27, 41, 64–65, 124, 154, 236, 287